SERIES 1 – FI[illegible]

ENGINEERING GRAPHICS PROBLEMS

HENRY CECIL SPENCER
Late Professor Emeritus of Technical Drawing
Formerly Director of Department
Illinois Institute of Technology

IVAN LEROY HILL
Professor Emeritus of Engineering Graphics
Formerly Chairman of Department
Illinois Institute of Technology

ROBERT OLIN LOVING
Professor Emeritus of Engineering Graphics
Formerly Chairman of Department
Illinois Institute of Technology

JOHN THOMAS DYGDON
Professor of Engineering Graphics,
Chairman of the Department,
and Director of the Division of Academic Services
and Office of Educational Services
Illinois Institute of Technology

JAMES E. NOVAK
Associate Director/Executive Officer,
Office of Educational Services
Illinois Institute of Technology

Macmillan Publishing Company
NEW YORK
Maxwell Macmillan Canada
TORONTO
Maxwell Macmillan International
NEW YORK OXFORD SINGAPORE SYDNEY

Printed in the United States of America.

Laser typesetting by *Ewing Systems*, 409 W. 24th St., Suite #14, New York, NY 10011.

Macmillan Publishing Company
866 Third Avenue, New York, New York 10022

Macmillan Publishing Company is part of the Maxwell Communication Group of Companies.

Maxwell Macmillan Canada, Inc.
1200 Eglinton Avenue East
Suite 200
Don Mills, Ontario M3C 3N1

ISBN 0-02-414940-3

Printing: 1 2 3 4 5 6 7 8 Year: 3 4 5 6 7 8 9 0 1 2

Preface

Engineering Graphics Problems–Series 1 is intended primarily for use with *Engineering Graphics* by Giesecke, Mitchell, Spencer, Hill, Loving, Dygdon, and Novak. All references and illustrations refer to that text, and the coverage is similarly divided into three areas: Part One, "Technical Drawing"; Part Two, "Descriptive Geometry"; and Part Three, "Graphs, Diagrams, and Graphical Computation." However, this workbook may be used with other reference texts.

In the interest of more efficient use of time, the objective has been to produce in as few sheets as possible a complete coverage of the fundamentals. Therefore, limited space is devoted to lettering and other exercises intended primarily to develop skills. Most of the problems are based upon actual industrial designs, and their presentations are in accord with the latest ANSI Y14 American National Standard Drafting Manual and other relevant ANSI standards. Moreover, a special effort has been made to present problems that are thought provoking rather than requiring a great deal of routine drafting.

An outstanding feature of this revision is that fractional-inch dimensions have largely been replaced by the decimal-inch and metric dimensions now used extensively in industry. A decimal and millimeter equivalents table and appropriate full-size and half-size scales are provided inside the front and back covers for the student's convenience.

All sheets are 8.5" × 11.0" in conformity with American National Standards for engineering drawing practices, a size that facilitates handling and filing by the student and the instructor. Several problems are printed on vellum to provide experience in the manner of commercial practice. It is expected that in most cases the instructor will supplement the worksheet problems with assignments from the text, to be drawn on blank paper. Many of the text problems are designed for Size A4 or Size A sheets, the same size as the easily filed worksheets. A supply of blank sheets and cross-section sheets, both rectangular and isometric, is provided on the reverse sides of the worksheets. Considerable emphasis is given to technical sketching. Numerous problems in this book make use of grids similar to the various cross-section papers available commercially.

The notation requirements for Part Two have been simplified to save time and to focus the student's attention on the solution of the spatial problems. The problems in this section are designed to give normal coverage in the area of descriptive geometry.

In line with the increased interest in charts, graphs, and graphical computation, a number of problems in these areas have been provided, including nomography, empirical equations, and graphical calculus. The basic elements of the design process and several suggested design problems are presented in Chapter 16, "Design and Working Drawings," of the text *Engineering Graphics.*

In response to the increased usage of computer technology for drafting and design, a number of problem sheets in computer-aided drafting (CAD) have been included. The problem sheets on detail drawings are presented to provide practice in making regular working drawings of the type used in industry. These are suggested for solution either by a computer-aided drafting system or by traditional drafting methods.

The instructions provide detailed information together with references to the text for each problem. The student is urged to study these instructions and references carefully before starting each problem.

The authors wish to express appreciation to their colleagues for many valuable suggestions and to numerous industrial firms who have so generously cooperated in supplying problem material. Comments and criticisms from users of this workbook will be most welcome.

Ivan Leroy Hill
Clearwater, FL

Robert O. Loving
Evergreen Park, IL

John Thomas Dygdon
Illinois Institute of Technology
Chicago, IL

James E. Novak
Illinois Institute of Technology
Chicago, IL

Contents

Worksheets

Part One. Technical Drawing and Design

Part Two. Descriptive Geometry

Part Three. Graphs, Diagrams, and Graphical Computation

Vellums

Printed on vellum, the following sheets appear at the back of this workbook.

Instructions

References are to Fifth Edition of *Engineering Graphics* (1993)
by Giesecke, Mitchell, Spencer, Hill, Loving, Dygdon, and Novak.

Throughout this workbook alternative dimensions, often not the *exact* equivalents, are given in millimeters and inches. Although it is understood that 25.4 mm = 1.00", it is more practical to use approximate equivalents such as 25 mm for 1.00"; 12.5 or 12 mm for 0.50"; 6 mm for 0.25"; 3 mm for 0.12"; etc. Exact equivalents should be used when accurate fit or critical strength is involved.

Part One. Technical Drawing

In general it will be found that the following leads are suitable for mechanical drawing: a 4H for construction lines and guide lines for lettering, a 2H for center lines, section lines, dimension lines, and extension lines, and an F for general line work and lettering. All construction lines on problems should be made *lightly* and *should not be erased.*

Drawing 1. Vertical Capitals and Numerals. References: §§4.1, 4.3, 4.5–4.18. Using an HB lead, letter the indicated characters in the spaces provided. These large letters and numerals may be sketched lightly first, and then corrected where necessary before being made heavy with the strokes shown. Omit the numbers and arrows from your letters. All lettering must be *clean-cut* and **black**. In the title strip, under DRAWN BY, draw light guide lines from the starting marks shown, plus random vertical guide lines, and letter your name with the last name first. Under FILE NO. letter identification symbol as assigned by your instructor.

Drawing 1A. Vertical Lowercase Lettering. References: §§4.21, 4.22, 4.24. In the upper half of the sheet, fill in the letters in the spaces provided, using an HB or F lead. Use the strokes shown, but omit the numbers and arrows from your letters. In the lower half of the sheet, draw vertical guide lines and then letter each line of lettering twice on the guide lines provided, using a sharp F lead. Make all letters *clean-cut* and **black**.

Drawing 2. Vertical Capitals and Numerals. References: §§4.1, 4.3, 4.5–4.18, 4.24. First draw light vertical guide lines at random from bottom to top of the sheet. Do not draw separate vertical guide lines for each line of lettering. Reproduce the lettering as exactly as you can, using an HB lead for the larger letters and a sharp F lead for the smaller letters. Note that the last line of lettering is to be lettered twice. All letters must be *clean-cut* and **black**.

Drawing 3. Vertical Capitals and Numerals. References: §§4.10, 4.15–4.18, 4.20, 4.24, 13.5, 13.7–13.11, 13.13–13.15, 13.17. On the left side of the sheet are shown a number of lettering applications. On the right, reproduce the lettering, arrowheads, and finish marks, using a sharp F lead. Except for the title TOOL HOLDER at the bottom, all lettering on the sheet is 3 mm or .12" high. Draw all guide lines with the aid of a lettering triangle or the Ames Lettering Guide, using a 4H or 6H lead. Letter, in vertical capitals, the title TOOL HOLDER, etc., on center as shown in Fig. 4.36 (b). Use height and spacing as specified. Underline TOOL HOLDER. All lettering must be *clean-cut* and **black**.

Drawing 4. Inclined Capitals and Numerals. References: §§4.1, 4.3, 4.5–4.17, 4.19. Using an HB lead, letter the indicated characters in the spaces provided. These large letters and numerals may be sketched lightly first, and then corrected as necessary before being made heavy with the stokes shown. Omit the numbers and arrows from your letters. All lettering must be *clean-cut* and **black**. In the title strip, under DRAWN BY, draw light guide lines from the starting marks shown, and letter your name with the last name first. Under FILE NO. letter identification symbol as assigned by your instructor.

Drawing 4A. Inclined Lowercase Lettering. References: §§4.21, 4.23, 4.24. In the upper half of the sheet, fill in the letters in the spaces provided, using an HB or F lead. Use the strokes shown, but omit the numbers and arrows from your letters. In the lower half of the sheet, draw inclined guide lines and then letter each line of lettering twice on the guide lines provided, using a sharp F lead. Make all letters *clean-cut* and **black**.

Drawing 5. Inclined Capitals and Numerals. References: §§4.1, 4.3, 4.5–4.17, 4.19, 4.20, 4.24. First, draw light inclined guide lines at random from bottom to top of the sheet. Do not draw separate inclined guide lines for each line of lettering. Reproduce the lettering as exactly as you can, using an HB lead for the larger letters, and a sharp F lead for the smaller letters. Note that the last line of lettering is to be lettered twice. All letters must be *clean-cut* and **black**.

Drawing 6. Inclined Capitals and Numerals. References: §§4.1, 4.3, 4.5–4.17, 4.19, 4.20, 4.24, 13.5, 13.7–13.11, 13.13–13.15, 13.17. On the left side of the sheet are shown a number of lettering applications. On the right, reproduce the lettering, arrowheads, and finish marks, using a sharp F lead. Except for the title TOOL HOLDER at the bottom, all lettering on the sheet is 3 mm or .12" high. Draw all guide lines with the aid of a lettering triangle or the Ames Lettering Guide, using a 4H or 6H lead. Letter the title TOOL HOLDER, etc., on center as shown in Fig. 4.36 (b). All lettering must be *clean-cut* and **black**.

Drawing 7. Conventional Lines. References: §§2.1–2.17, 2.24, 2.25, 2.33, 2.43, 2.46, 13.7, 17.1–17.6.

Space 1. Using the T-square, complete the conventional lines to fill the right half of the space, and to match the lines in the left half. Use a sharp F lead with a conical point, Fig. 2.8 (a), for the visible line, hidden line, cutting-plane line, short-break line and stitch line; a 2H lead for the center line, section line, dimension and extension line, longbreak line, and phantom line. Make all lines black, but of the correct widths.

Space 2. Draw the *view* of the Anchor Slide full size, locating the view by the starting point indicated. Your final lines should correspond to those given in Space 1, with three distinct thicknesses of lines. Omit dimensions unless assigned. For section lining see §9.4.

Drawing 8. Use of T-square and Triangles. References: §§2.5–2.17, 2.20–2.25, 2.32, 2.46. Draw lines to fill the spaces as follows.

Space 1. Set of four horizontal visible lines 20 mm apart and centered in the working space. First, locate the center of the space by drawing light diagonals (draw only light dashes crossing at the center). Through this point draw a construction line that will be at right angles to the required lines. Along this construction line set off 20 mm spaces. Use scale in the position showing in Fig. 2.54, Step III, and draw the required lines through these points at right angles to the construction lines as shown in Fig. 2.20.

Space 2. Set off five vertical hidden lines, 20 mm apart and centered in the working space. Locate the center of the space as for Space 1 and set off 20 mm spaces along a construction line through the center and perpendicular to the required lines. Use scale in position shown in Fig. 2.54, Step II, and draw lines as shown in Fig. 2.21.

Space 3. Inclined section lines 30° with horizontal, sloping upward to the right, and spaced 20 mm apart. Find the center of the space as for Space 1. Along a construction line through the center and perpendicular to the required lines, set off 20 mm spaces, making one mark at the center. Through these points draw the required lines. See Fig. 2.26 E and L.

Space 4. Inclined center lines 75° with horizontal, sloping downward to the right, and spaced 20 mm apart. Locate the center of the space as before, draw construction line at right angles to required lines, and set off distances on this line, making one mark at the center. Then draw the required lines at right angles to this line through these points. See Fig. 2.23 F and M.

Space 5. Draw cutting-plane lines parallel to given line and at 20 mm intervals to fill the space. Along a construction line perpendicular to the given line, set off 20 mm spaces and through these points draw the required lines.

Space 6. Draw visible lines 20 mm apart and perpendicular to the given line. Arrange so that one visible line passes through the center of space (some division points will fall outside the rectangle). First locate the center of the space as before. Along a construction line through this center and parallel to the given line set off 20 mm spaces. Draw the required lines through these points and perpendicular to the given line.

Drawing 9. Scale. References: §§2.4–3.2.

Space 1. Use architects, engineers, or metric scales as necessary. Measure lines A, B, C, E, F, G and J at the scales shown, and indicate the scaled lengths (L) at the right. At D, H, I, and K, draw lines of specified lengths at scales shown. Terminate the lines in the same manner as for given lines. At L through N, determine the scales and lengths of lines, and record the scales and lengths in the spaces provided. Make lettering the same as that given. Draw vertical guide lines and additional horizontal guide lines as needed for your lettering.

Line L is over 500' and under 600' in length.
Line M is between 625 m and 650 m in length.
Line N is one twenty-fourth size.

Spaces 2–9. Regard the partially dimensioned layout in each space as a portion of an actual drawing. All given dimensions are in inches for Spaces 2, 3, 5, and 7. Spaces 4, 6, 8, and 9 are given in millimeters. A scale on the drawing represents the ratio of the size on the drawing to the actual size of the object as $1/2$, $1/10$, etc. This ratio can be expressed either in inches or millimeters. Determine the scale used for the drawing and indicate it in the spaces provided. Draw vertical guide lines as needed for your lettering.

Drawing 10. Geometric Constructions. References: §§5.1–5.6. Show light construction on all problems. Add center lines where necessary.

Space 1. Reference §§5.8, 5.9. Locate and draw hole as indicated. Add center lines.

Space 2. References: Fig. 5.28 (a). Complete the view of the Special Washer.

Space 3. References §§5.14, 5.15. Complete the view of the Rack. Start the first tooth at A as indicated.

Space 4. Reference §5.18. Locate centers for holes as specified. Draw holes and add center lines.

Space 5. Reference: §5.19. Complete the view of the End Guide by adding the 90° tip as indicated in the given specifications.

Space 6. References: §5.36, Fig 13.42 (b). Complete the view of the Bracket.

Drawing 11. Geometric Constructions. References: §§5.1–5.6. Show light construction on all problems. Add center lines where necessary.

Space 1. References §§5.25, 5.35. Complete the view of the Lever.

Space 2. Reference: §5.39. Complete the view of the Clamp.

Space 3. References:§§5.36, 5.41. Complete the view of the Bracket.

Space 4. Reference: §5.40. Complete the view of the Link.

Space 5. References: §§5.40, 5.41. Complete the view of the Toggle Link.

Space 6. Reference: §5.57. Draw the approximate ellipse as indicated to complete the view of the Bearing.

Drawing 12. Geometric Constructions. Show light construction on all problems.

Space 1. References §§2.54, 5.51. Using the concentric-circle method, draw the profile of the 72 mm × 102 mm elliptical cam, starting at point A. Use a minimum of 16 points to establish the ellipse. Draw the ellipse with the aid of the irregular curve.

Space 2. References: §§2.54, 5.58. Using the method of Fig. 5.57 (c), design the parabolic gateway arch as specified. Draw the parabola with the aid of the irregular curve.

Space 3. References: §§2.54, 5.59. Using the method of Fig. 5.59 (b), design the parabolic curve between points X and Y and tangent to XO and YO as specified. Draw the parabola with the aid of the irregular curve.

Space 4. References: §§2.54, 5.63. Using the method of Fig. 5.63 (b), construct the right-hand helix as specified. Divide the semicircle of the base into 15° intervals for the equivalent of 24 points for a complete circle. Show visibility and draw the helix with the aid of the irregular curve.

Drawing 13. Multiview Technical Sketching. References: §§6.1–6.10, 6.18–6.31. Sketch the views as indicated. Make all final lines *clean-cut* and **black**. In the space provided, carefully letter the names of the necessary views.

Drawing 14. Multiview and Isometric Technical Sketching. References: §§6.1–6.31. Sketch the views as indicated, spacing the views four squares apart. Make all final lines *clean-cut* and **black** so the sketches will stand out from the grid lines.

Drawing 15. Missing Lines. References: §§6.5, 6.11–6.20, 6.25–6.31. Each problem is incomplete because lines are missing from one or more views. Add all missing lines freehand, including center lines.

Drawing 16. Multiview Technical Sketching. References: §§6.5–6.10, 6.18–6.31. Sketch views of the Holder Push Dog as indicated. The counterbored hole, Fig. 7.40 (c), and the two holes in the base are *through* holes. Make all final lines *clean-cut* and **black** so sketches will stand out from the grid lines.

Drawing 17. Missing Views. References: §§6.25, 6.26, 7.1–7.14. In each problem two complete views are given, and a third view is missing. Add the third view in each case freehand. Make all final lines *clean-cut* and **black** so the sketches will stand out from the grid lines.

Drawing 18. Missing Views. References: §§6.25, 6.26, 7.1–7.30, 7.32, 7.33. In each problem two views are given. Add the third view as specified, using instruments. The given views are complete.

Drawing 19. Missing Views. References: §§7.1–7.30, 7.33. In each problem two complete views are given. Add the third view in each case, using instruments. For Prob. 1, study Fig. 7.35.

Drawing 20. Missing Views. References: §§7.1–7.30, 7.33. In each problem two complete views are given. Add the third view in each case, using instruments.

Drawing 21. Missing Views. References §§7.1–7.30, 7.33. In each problem two complete views are given. Add the third view in each case, using instruments.

Drawing 22. Missing View. References: §§7.1–7.30, 7.33–7.36, 13.17. Two complete views are given of the Support Bracket. Add the top view, using instruments. The small fillets and rounds are 2.5 mm R and may be drawn freehand. If assigned, show finish marks for all three views.

Drawing 23. Sectional Views. References: §§9.1–9.7, 9.15, 12.1–12.7. Sketch the sections as indicated. Make section lines *thin* to contrast well with heavy visible lines. Make all final lines *clean-cut* and **black** so the sketches will stand out from the grid lines.

Drawing 24. Sectional Views. References: §§2.11, 7.34, 9.5, 9.8, 9.9, 9.11–9.13, 9.15. Draw the indicated sectional views, using instruments. In all four problems, unspecified fillets and rounds are 1.5 mm R, to be drawn freehand. In Space 2, draw revolved section in the opening in the top view, and show break lines on each side of the section. See Fig. 9.18. Draw break line from A to B in the front view as in Fig 9.35 (d); then draw broken-out section to the right of the break line. Label the cutting plane and the section in Space 3.

Drawing 25. Sectional Views. References: §§7.34, 9.1–9.6, 9.10, 12.5, 13.17. Draw the indicated sectional views, using instruments. In Prob. 1, add finish marks (except for holes). In Prob. 2, show all visible lines behind the cutting plane in each section.

Drawing 26. Sectional Views. References: §§9.1–9.6, 16.21. Draw the indicated sectional views, using instruments. If assigned, add finish marks to all views in Space 1. In Space 2 is shown a portion of an assembly in full section with a round shaft extending through a cast-iron cover and a steel plate which are held together by bolts. Section line the sectioned areas, using symbolic section lining for each material, or if assigned, use the general use symbol for all sectioned areas.

Drawing 27. Sectional Views. References: §§7.33, 7.34, 9.13, 9.15, 15.9. Draw Section A–A and Section B–B as indicated. If assigned, add finish marks to given and sectional views.

Drawing 28. Primary Auxiliary Views. References: §§10.1–10.10. Sketch auxiliary views as indicated. Make visible lines and hidden lines **black** so that the views will stand out clearly from the grids. Letter folding lines as shown in Fig 10.3, and reference planes as in Fig. 10.6. Use folding lines in Spaces 1 and 2, and reference plane lines in Space 3. Use either folding lines or reference plane lines in Space 4 as assigned. In Spaces 2, 3, and 4, include all hidden lines.

Drawing 29. Primary Auxiliary Views. References: §§10.1–10.10, 10.14, 10.16, 10.17. Add, by freehand sketching, any missing lines in the regular views or auxiliary views.

Drawing 30. Primary Auxiliary Views. References: §§10.1–10.12, 13.14. Draw the required auxiliary views, using instruments. In Space 1 add the numbers in the auxiliary view, making them the same height as those given. In the remaining problems use numbers if needed, or if assigned. Show folding lines or reference planes in all problems, as assigned. In Space 4, surface A is an inclined surface, §7.21, and surface B is an oblique surface, §7.23. Dimension the required angle in degrees, §13.14. Include all hidden lines in all problems.

Drawing 31. Offset Auxiliary Section. References: §§7.34, 9.1–9.6, 9.11, 10.17, 12.5, 12.20, 15.15, 15.16. Draw the indicated offset auxiliary section. If assigned, add finish marks to all views.

Drawing 32. Secondary Auxiliary Views. References: §§10.11, 10.19–10.22. Draw the required views as indicated, including all hidden lines (except in the partial auxiliary view in Space 1). In Space 2 measure the true angle between surfaces A and B with a protractor, and dimension the angle in degrees, Fig. 13.17 (e). Use either reference-plane lines or folding lines as assigned. In Space 2 space the primary auxiliary view approximately 30 mm from the given top view.

Drawing 33. Revolutions. References: §§11.1–11.11. Use the standard phantom line or alternate position line for revolved positions of all lines except center lines which remain unchanged.

Space 1. Revolve point **3** as required. Dimension the angle of revolution as in Fig. 11.6.

Space 2. Revolve line **3–4** as required in a manner similar to that used in View 1 of Fig. 11.9 (b) and indicate the angle of revolution. Add TL to the true-length view of the line 3–4.

Space 3. Revolve the prism as specified about edge **2–3** as the axis. Add TS to the true-size revolved view of the surface 1–2–3–4–5.

Drawing 34. Isometric Sketching. References: §§6.1–6.8, 6.12–6.14. Using an HB lead, sketch isometrics of the six given machine parts in spaces provided for each problem. Note the given starting point A. Omit hidden lines in all problems.

Drawing 35. Isometric Drawing. References: §§18.5–18.18. Omit hidden lines in all problems. All "box construction" and other construction lines should be made lightly with a sharp 4H lead and should not be erased. Darken all visible lines with a sharp F lead.

Spaces 1, 2, and 3. Draw isometric drawings, starting at corner A and using the dividers to transfer distances from the views to the isometrics.

Space 4. Draw isometric drawing, locating the corner A as indicated. Use the scale to set off dimensions. Do not transfer distances with dividers, as the given drawing is not to scale. Show construction for the 30° angle.

Space 5. Complete the isometric drawing, using the information supplied in the reduced-scale drawing. Unless otherwise instructed, use the approximate ellipse of §18.18.

Space 6. Complete the isometric drawing, transferring measurements directly from the given views to the isometric with dividers. Draw the final curves with the aid of the irregular curve, §2.54.

Drawing 36. Isometric Drawing. References: §§18.5–18.18. Draw isometric drawings, locating them at point A as indicated. Omit hidden lines. Make all construction lines *lightly* with a 4H lead, and do not erase. In particular, show complete construction for angles. Use a sharp F lead for the visible lines.

Drawing 37. Isometric Drawings. References: §§18.5–18.18. Draw isometric drawings, locating them at points A as indicated. Omit hidden lines. Make all construction lines *lightly* with a 4H lead, and do not erase. In particular, show complete construction for angles. Use a sharp F lead for the visible lines. In Space 1, part of the box construction will overlap the given views of the machine part. Note that the isometric in Prob. 2 is to be drawn with reversed axes, as shown in Fig. 18.10, and is to be half size.

Drawing 38. Oblique Projection. References: §§6.4, 6.5, 6.15, 6,16. Using the method of Fig. 6.28, make freehand drawings of the machine parts shown, using starting corners indicated. Omit hidden lines. Make visible lines **black** (HB lead) so the pictorial sketches will stand out clearly from the grids.

Drawing 39. Oblique Projection. References: §§19.1–19.6. Draw oblique pictorials full size with instruments, starting at points A and B as indicated. The angles of the receding lines are 30° or 45° with horizontal, as shown at these starting points. Omit all hidden lines. Show all construction, *especially* for the points of tangency. Study Fig. 19.13 carefully.

Drawing 40. Dimensioning. References: §§13.1–13.25, 13.30, 13.31. Use the complete decimal dimensioning system with metric values. If assigned, use decimal-inch equivalents. Add dimensions freehand, spacing dimension lines approximately 10 mm from the views and 10 mm apart. Include necessary finish marks in Prob. 2. Note that in Prob. 2 the drawing is half the size of the actual part. Dimension the keyway in the manner shown in Fig. 13.44 (x). The two small holes are drilled, Appendix 11, and the large hole is bored. In the bored-hole note, specify the diameter to two decimal places.

Drawing 41. Dimensioning. References: §§13.14–13.25, 13.30, 13.31. Use the complete decimal dimensioning system with metric values. If assigned, use decimal-inch equivalents. Add dimensions mechanically, spacing dimension lines 10 mm from the views and 10 mm apart.

Space 1. Since the material is CRS (cold rolled steel), the object is understood to be finished all over, and no finish marks are necessary. In this problem the small hole is drilled, and the two medium-sized holes are drilled and countersunk. See Fig. 13.44 (c) and Appendix 11. The large hole is reamed. In the note indicate this diameter in decimals to two places.

Space 2. Add finish marks. The small holes are drilled and equally spaced. The central portion is cored, and the large end holes are bored. Use the note 19.05–19.10 BORE. All fillets and rounds are 3R.

Drawing 42. Mating Parts Dimensioning. References: §§13.9, 13.20–13.27, 13.31, 14.1–14.8. Dimension the Base fully, including finish marks and notes. Space dimension lines 10 mm from the views and 10 mm apart. Use the complete decimal dimensioning system with metric values except for certain standard parts. If assigned, use decimal-inch equivalents. The large hole is to be reamed to metric values equivalent to an RC 6 fit (see Appendix 5). Note that the RC 6 fit also applies to the diameter of the Special Bolt on Drawing 43. The two smaller holes are to be drilled with 0.5 mm allowance and spotfaced for M8 × 1.25 hexagon head cap screws.

Drawing 43. Mating Parts Dimensioning. References: §§13.9, 13.20–13.27, 13.31, 14.1–14.8, 15.21. These three parts fit with the Base on Drawing 42 and are to be dimensioned fully in the manner used on Drawing 42. Dimension the length of the Bushing with an allowance of 0.08 mm and a tolerance of 0.05 mm figured from the maximum length as the basic size. The tolerances should permit the Bushing to turn freely on the Special Bolt.

If assigned, draw an assembly of the Roller Guide with part identification numbers and a parts list. See the figure for Hydraulic Grease Fitting details. References: §§16.14, 16.20–16.22.

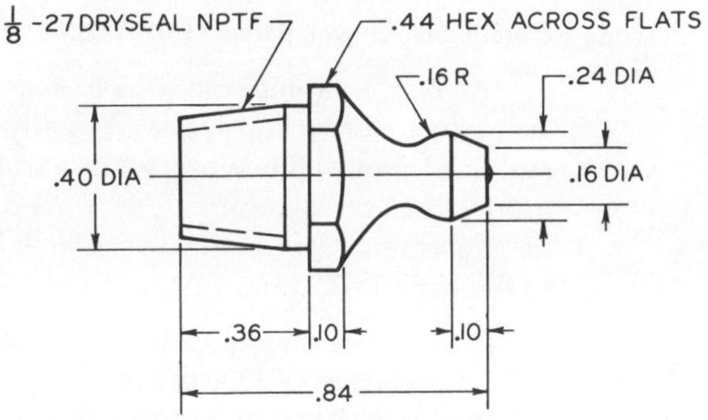

1/8-27 DRYSEAL NPTF
.44 HEX ACROSS FLATS
.16 R
.24 DIA
.40 DIA
.16 DIA
.36
.10
.10
.84

NO. 8585 HYDRAULIC GREASE FITTING
STEEL-BRIGHT ZINC PLATE

Drawing 44. Threads and Fasteners. References: §§15.1–15.24, 15.29–15.35. Draw light guide lines from the marks indicated, and letter the answers in the spaces provided. Standard abbreviations may be used to avoid crowding. See Appendix 4.

Drawing 45. Unified Threads. References: §§15.3–15.10, 15.20, 15.21.

Piston Rod. Complete the partial front view and the end view by drawing the threads specified.

Gland. Add the specified threads to the given views. Note that the front view is a half section, and the incomplete side view is a half view. Omit hidden lines in the sectioned front view, and add only the lines necessary to show the thread in the half side view. Complete the section lining in the sectional view.

Adjustable Link. This is an assembly of three parts, as indicated. The Link Base is threaded through. The Eye Bolt is engaged to the depth indicated, and its position locked by the Jam Nut. Complete the assembly view and also the half left-side view of the Link Base. Omit the other parts of the assembly in the half view. Complete all section lining in the sectional view.

In all three problems, complete the thread-note leaders.

Drawing 46. Detailed Acme and Square Threads. References: §§15.3–15.7, 15.12, 15.13, 15.15, 15.21.

Adjusting Screw. Draw the specified Acme threads to complete the view. Complete the leaders and add arrowheads touching the threads. Construct the threads so as to be symmetrical about the central neck of the screw.

Leveling Jack. Draw the specified Square threads to complete the assembly view. Note that the scale of the drawing is double size. Add necessary section lining and complete the thread-note leader.

Drawing 47. Threads and Fasteners. References: §§15.7–15.10, 15.23–15.28; Appendix 27. Draw specified threads and fastener details, using the schematic thread symbols unless otherwise assigned. Complete the section lining and leaders where required. Chamfer ends of threads 45° × thread depth in Probs. 1, 2, and 3.

Part Two. Descriptive Geometry

Accuracy. Graphic solutions to space problems require accurate measurements and clean, sharp line work. Properly sharpened F and 2H leads will be found suitable for most line work, while a sharp 4H lead is preferred for construction and guide lines for lettering. The F lead is normally used for lettering. Make all measurements from center to center of lines and when no scale is specified, measure to the nearest 0.2 mm.

Unless otherwise specified, the basic unit for the given scale is the millimeter. Hence, a scale of 1/2000 is 1 mm to 2000 mm or 2 m. For measurements equal to or greater than 1000 mm, indicate these measurements in meters.

Notation. A certain amount of lettering is necessary on all graphical solutions. A minimum amount of notation should include the following:

1. Label at least one point in each view, or label points that are mentioned in the instructions.
2. Label all folding lines employed.
3. Show the symbols for EV (edge view), TL (true length), TS (true size), and LI (line of intersection) when they are a part of the solution.
4. Show given and required information such as angles, distances, bearings, and other numerical items *on the views* where measured or set off.

Drawing 48. Visibility of Lines. Reference: §21.1. Use standard alphabet of lines, §2.11, to complete the solutions. Change the dotted lines to standard lines in Spaces 3 and 4.

Drawing 49. Points and Lines. References: §§7.15, 21.1, 21.2.
Space 1. Point 2 is 33 mm to the right of point 1, 25 mm below point 1, and 20 mm in front of point 1. Show these dimensions on the drawing.
Space 2. Line 1–2 is 38 mm long (2 is behind 1). Line 1–3 is a 40 mm long frontal line. Show these dimensions on the drawing. Line 2–3 is a profile line. Dimension the true length of line 2–3.
Space 3. Line 1–2 is 33 mm long. Show this on the drawing. The front view of line 2–3 is true length as indicated.
Space 4. Point 5 is one line 1–2 and is 15 mm above point 1. Point 6 is on line 3–4. Line 5–6 is a horizontal line. Dimension the length of line 5–6.
Space 5. Note that point 4 is to be moved vertically *in space.*
Space 6. Triangle 1–2–3 is the base of a pyramid. Vertex V is 5 mm behind point 1, 10 mm to the left of point 2, and 28 mm above point 3. Show these dimensions on the drawing.

Drawing 50. True Length of Line. References: §§10.18, 21.3. Use auxiliary views. Dimension the required angles in degrees as in §13.14, and also dimension the true lengths in millimeters for Spaces 1, 2, and 3. Show all given data on the drawing for Space 4.

Drawing 51. True Length, Bearing, and Grade. References: §§21.3, 21.5. Use auxiliary views. For all problems, indicate the percent grade on the drawing, as in Figure 21.14 (b). For Spaces 1, 2, and 3, also show the numerical values of the bearing and true length on the drawing.

Drawing 52. True Length by Revolution. References: §§11.10, 21.4. Use American National Standard phantom lines for lines in alternate or revolved-position views.
Space 1. Dimension the true length, slope, and bearing on the drawing.
Space 2. Dimension ∠F, ∠P, and the true length on the drawing.
Space 3. Show the 30° angle on the drawing.
Space 4. Show all given data on the drawing.

Drawing 53. Point View of Line. References: §§10.11, 21.6.
Space 1. Include the minimum notation as recommended prior to Drawing 48 above.
Space 2. Dimension the clearance on the drawing.
Space 3. Dimension the true distances for comparison.

Drawing 54. Points and Lines in Planes. References: §§21.7, 21.8.
Space 1. Indicate your answer in the space provided.
Space 2. The *top* view of point 4 and the *front* view of point 5 are shown.
Space 4. See also §21.1. Show given data on the drawing.

Drawing 55. True Size of Planes. References: §§7.15, 10.1, 10.21, 21.9, 21.10.
Space 1. Indicate values for calculation of the area on the drawing and show your calculations. Record the area to two significant figures only, in the space provided.
Space 2. See Appendix 36 for information on pipe fittings.

Drawing 56. True Size of Planes. References: §§7.15, 10.1, 10.21, 21.9, 21.10. If assigned, revolution may be used instead of a secondary auxiliary view, §§11.11, 21.10.
Space 1. If assigned, plot the views of the circle in the given views. See §5.11.
Space 2. Show usage of given data on the drawing.

Drawing 57. Piercing Points. Reference: §21.11. Use the edge view method on this sheet and show proper visibility in all views. Show EV where appropriate and encircle the piercing points in all views.

Space 3. Do not show a hidden line for that segment of line 1–2 that is within the pyramid.

Drawing 58. Piercing Points. Reference: §21.11. Use the cutting-plane method on this sheet and show proper visibility in all views. Show EV on the lines as edge-view cutting planes and encircle the views of the piercing points.

Space 3. Do not show a hidden line for that segment of line 1–2 which is completely within the pyramid. However, include any extension lines necessary to clarify the method of solution.

Space 4. Note that the given views are a front and a primary auxiliary view.

Drawing 59. Intersection of Planes. Reference: §21.12. Show the symbol EV wherever appropriate and label the line of intersection with the symbol LI.

Space 1 and 2. Show complete visibility.

Space 3 and 4. Visibility is of no concern in these problems, but special care is needed to insure accuracy of your solutions.

Drawing 60. Dihedral Angles. References: §§10.11, 21.9.

Space 1. See also §5.7. Dimension the angles on the drawing. Show complete visibility.

Space 2. See also §21.12. Special care is needed to assure accuracy in your solution of this problem. Show the angle on your drawing. Show complete visibility.

Drawing 61. Angle Between Line and Plane. Reference: §21.13.

Space 1. Dimension the angle on the drawing.

Space 2. This is an "open ended" problem (as are most engineering problems) in that there are many possible answers. Show use of given data on the drawing.

Drawing 62. Parallel Lines and Planes. References: §§7.25, 21.8, 22.1, 22.3, 22.4.

Space 1. Demonstrate your answer graphically and indicate it by a check mark(s) (✓) in the space provided.

Space 2. You can check vertical alignment with a T-square and a triangle, but do not perform any actual constructions. Indicate your answer(s) by a check mark (✓) in the spaces provided.

Space 3 and 4. Use only the given views.

Drawing 63. Parallel Lines and Planes. References: §§7.25, 21.5, 22.1–22.4. Show bearing on the drawing in Space 1.

Drawing 64. Perpendicular Lines. References: §§22.5, 22.6. These problems are to be solved without the construction of auxiliary views. Use the symbol as shown in Fig. 5.1(k) to show lines drawn perpendicular.

Drawing 65. Perpendicular Lines and Planes. References: §§22.5–22.7.

Space 1. Do not construct additional views.

Space 2. Use an auxiliary view, or only the given views, or both, as assigned. Show the length of the altitude on the drawing.

Space 3. Use auxiliary views. Show *all* views in completed form with proper visibility.

Drawing 66. Common Perpendicular. References: §§22.5, 22.9. These problems are designed to be solved by the point-view method. Use of the plane method is not recommended because of space limitations.

Space 1. Show answers on views where measured. Be sure to check accuracy with divider distances.

Space 2. Indicate the actual clearance distances on the drawing where measured. Show the front and side views of lines representing the two distances. Indicate with a check mark (✓) in the spaces provided, the answer to the matter of inspection.

Drawing 67. Shortest Connector at Specified Angle. References: §§22.10, 22.11.

Space 1 and 2. Show given data on the drawing where measured.

Drawing 68. Intersections of Planes and Polyhedra. References: §§23.1, 23.2.

Space 1. The edge-view method is suggested. Be sure to check accuracy with divider distances. Show visibility.

Space 2. Use the cutting-plane method, including the EV symbols. Show complete visibility except that hidden line segments of the plane's boundaries within the prism should be omitted.

Drawing 69. Intersections of Planes and Curved Surfaces. Reference: §23.3

Space 2. Use cutting planes that cut circles from the cone.

Drawing 70. Intersections of Prisms and Pyramids. References: §§23.4–23.6. Omit hidden line segments of lateral edges of either solid which are within the other solid.

If assigned: On a separate sheet develop the surface of one or more of the solids, including the intersections. References: §§24.1–24.3, 24.6, 24.7.

Drawing 71. Intersections of Circular Forms. Reference: §23.8. Determine the figures of intersection and show visibility.

Drawing 72. Intersection of Oblique Cone and Cylinder. Reference: §23.9. Use a minimum of five cutting planes to establish the figure of intersection. Show visibility.

Drawing 73. Parallel-Line Development. References: §§24.1–24.5. Show full developments inside up. Omit the bases.

Space 2. Calculate the length of the development for the cylinder. Start development with the shortest element.

Drawing 74. Radial-Line Development. References: §§24.1, 24.6–24.9. Show developments inside up. Start with shortest seam. Omit bases.

Drawing 75. Triangulation. References: §§24.10, 24.12, 24.13. Show half development inside up. Omit the cylindrical and prismatic connectors at the top and bottom.

Drawing 76. Line and Plane Tangencies. References: §§25.1–25.4. If assigned, where alternative solutions are possible, show the second solution with a phantom line.

Drawing 77. Lines and Planes at Specified Angles. Reference: §25.5. If assigned, show alternative solutions with phantom lines.

Drawing 78. Topographic Mapping. References: §§26.1, 26.2. Using 10 m contour intervals, plot by interpolation the contour lines. The graphical method of §5.15 is suggested. Draw profiles at grids 5 and 7 (or any others as assigned) in the space below the map. Use a vertical scale of 1/2000, and use the profiles to check graphically the previously interpolated points where the contours cross these grids.

Drawing 79. Mining and Geology—Strike, Dip, and Thickness. References: §§26.3, 26.4. Show all numerical data, required or given, on the drawings.
Space 4. See §25.5.

Drawing 80. Mining and Geology—Outcrop. References: §§26.3–26.5.
Part a. Show the numerical answers on the drawing.
Part b. Show all projection lines (light construction) between views and use a dash or dash-dot coding for the outcrop lines. The outcrop lines should be heavy lines. Colored pencil may be used. Shade the outcrop area.

Drawing 81. Civil Engineering—Cut and Fill. Reference: §26.6. Show light construction lines for all plotted points. Crosshatch the cut and fill areas in opposite directions and also crosshatch the given rectangles for the KEY. Do not crosshatch the roadway area.

Drawing 82. Spherical Triangle. Reference: §26.7. Dimension and identify on the drawing all given and required angles. If assigned, plot the elliptical arcs on the views.

Drawing 83. Spherical Triangle in Navigation. Reference: §26.8. Dimension and identify on the drawing all given and required angles. Show calculations for the required great-circle distance. If assigned, plot the front and top views of the course.

Drawing 84. Concurrent Coplanar Views. References: §§27.1, 27.2, 27.6. Show all scalar quantities on each VECTOR DIAGRAM.
Space 1. Use Bow's notation.
Space 2. Use Bow's notation without regard to the reactions R_1 and R_2, determining these later.

Drawing 85. Noncurrent Coplanar Vectors. Reference: §27.3. For the FORCE DIAGRAM, add Bow's notation externally to the space diagram. After completing the FORCE DIAGRAM, extend Bow's notation to the internal spaces of the SPACE DIAGRAM and construct the STRESS DIAGRAM. Label each vector with its scalar value.

Drawing 86. Resolution of Concurrent Noncoplanar Vectors. Reference: §27.5. Label the stress values along each true-length view.
Space 1. Since member O–1 appears in point view in the top view, the method of Fig 27.11 is recommended. Add Bow's notation on the top view after temporarily displacing O–1.
Space 2. The method of Fig. 27.12 is convenient here, with Bow's notation added in the view showing the plane of two unknown forces in edge view.

Part Three. Graphs, Diagrams, and Graphical Computation

Drawing 87. Pie and Bar Charts.
Space 1. References: §§28.20, 28.21. Worldwide production of cars, trucks, and buses for 1990 totaled 48,113,000 units, with the following percentage distribution.

Japan	28%
U. S.	21%
Germany	11%
France	8%
Canada	4%
Italy	4%
Russia	4%
United Kingdom	3%
All others	17%

Divide the pie chart to illustrate the given data. Label and show the value for each sector. Use 3 mm or 0.12" engineering lettering. Place the sum of the two largest sectors about a vertical center line in the lower portion of the circle. Centered below the title, letter 1990. In the lower right hand corner, letter TOTAL OUTPUT 48,113,000.

Space 2. References: §§28.16, 28.17. Construct a bar chart for the following data on U.S. nuclear power generation.

Nuclear Power Generation
(Millions of kilowatt-hours)

1978	275,000
1979	255,000
1980	250,000
1981	270,000
1982	285,000
1983	295,000
1984	330,000
1985 (Est.)	385,000

Use 12 mm wide vertical bars beginning at the given horizontal base line. Allow 6 mm spaces at the beginning and end and between the bars. Since the value for the bar for 1985 is estimated, place EST. above the top of the bar in 2.3 mm or 0.09" lettering. Shade the bars as in Fig. 28.31 (a). Title the chart in 3 mm or 0.12" letters. Draw horizontal grid lines for each 50,000 value indicated, but do not draw them across the bars.

Drawing 88. Engineering Graphs. References: §§28.3–28.9. Plot the following data on the rectangular coordinate grid in Space 1 and again on the semilogarithmic coordinate grid in Space 2. Plot TEMPERATURE (T) on the abscissa.

Rupture Strength of T.D. Nickel—
High Temperature Alloy

Temperature (T), °F	*100-hr Rupture Stress (s). 1000 psi (log scale)*
1200	24
1400	17
1600	12.5
1800	9
2000	6.5
2200	4.75
2400	3.5

Add titles to each graph: 100-HR RUPTURE STRESS OF T.D. NICKEL. Subtitle: HIGH TEMPERATURE ALLOY. Also add appropriate scale designations and captions. Show data points with small circles as in §28.6.

If assigned, find by the selected points method of §30.6 an empirical equation for the data plotted in Space 2.

Drawing 89. Parallel Scale and N-Chart Nomographs.

Space 1. References: §§29.8, 29.9. Fill in the blanks in the DESIGN TABLE and complete the alignment chart for the equation $V_f = V_0 + gt$, the velocity of a free-falling body, where

V_f = final velocity, ft/sec
V_0 = initial velocity, 0–40 ft/sec
t = time, 0–20 sec
g = acceleration due to gravity, constant = 32.2 ft/sec^2

Check the chart by substituting the values of $V_0 = 10$ and $t = 12$ in the equation. Show the isopleth thus located. Show title and equation above the chart. Add a KEY at an appropriate place on the chart.

Space 2. References: §§29.14, 29.15. Complete the N-chart for the equation $W = \frac{1}{2}LI^2$, the energy of a magnetic field, where

W = energy, joules
L = inductance 0–4 henries
I = current, 0–30 amperes

The design for the partially given N-chart is based on rewriting the equation as $\frac{1}{2}I^2 = {}^W\!/_L$, which makes the parallel scales uniform as shown, and places the nonuniform $\frac{1}{2}I^2$ function on the diagonal scale, which is always nonuniform. No design table is required, but fill in the DATA TABLE used for calibrating the diagonal scale. Add title and equation above the chart and a KEY at a convenient open area on the chart. Check the chart and show the isopleth for $L = 3$ and $I = 24$.

Drawing 90. Logarithmic Parallel-Scale Alignment Chart. References: §§29.10, 29.11. Also see §28.9. Fill in the blanks in the DESIGN TABLE and complete the parallel-scale alignment chart for the equation $\mu = f/N$ for the coefficient of friction, where

μ = coefficient of friction, 0.2–0.6
f = frictional force, lb
N = normal force of plane of movement, 5–150 lb

For the design of a logarithmic parallel-scale chart, the given equation is rewritten as

$$\log f = \log \mu + \log N$$

Check the chart and show the isopleth for the values $\mu = 0.5$ and $N = 10$. Show title and equation above the chart and add a KEY at a convenient open area on the chart.

Note: If log scales are not available, use the following values in connection with an engineers' scale.

log 1.5 = .176	log 4 = .602	log 7 = .845
log 2 = .301	log 5 = .699	log 8 = .903
log 3 = .477	log 6 = .778	log 9 = .954

Drawing 91. Empirical Equation—Rectangular Coordinates. References: §§30.1–30.4. Plot the following data on the given area with PRESSURE (P) on the abscissa and EFFICIENCY (V) on the ordinate.

Gear Pump Efficiency
(internal-external gear pump at 600 rpm)

Pressure (P), psi	*Volumetric Efficiency (V), %*
100	82.0
200	75.0
300	68.5
400	61.5
500	54.5
600	47.5
700	40.0
800	33.0

Show suitable title, scale captions, and designations. In the space above the graph show calculations for an empirical equation given by the slope-intercept method and also by the selected points method. Indicate on the graph the sources of all data used in the calculations. Use appropriate guide lines for all lettering.

Drawing 92. Empirical Equation—Log-Log Coordinates. Reference: §30.7. See also §28.9. Plot the following date with HEAD (H) on the abscissa and DISCHARGE (D) on the ordinate.

Head Discharge, 1000 m Length of 4" pipe

Head (H), m	*Discharge (D), liters/sec*
2	50
4	71
6	87
10	112
15	140
20	160
25	180
30	195
40	225

Show suitable title, scale captions, etc. In the space below the graph show calculations for an empirical equation by the slope-intercept method and also by the selected points method. Indicate on the graph the sources of all data used in the calculations. Use appropriate guide lines for all lettering.

Drawing 93. Graphical Algebraic Solutions. Reference: §31.1

(1) Plot the two given equations and obtain their simultaneous solutions to three significant figures. It will be necessary to calculate the coordinates of points relatively close together in the vicinity of the solutions in order to obtain the required accuracy.

(2) Determine the X-roots of the second equation, also to three significant figures. For all solutions, substitute the values in the equations and compute the results to demonstrate the accuracy obtained. Show these calculations at some convenient open area on the graph. Label the solutions and the curves.

Drawing 94. Graphical Differentiation. References: §§31.2–31.6. The given curve shows the results of a speed test of an automobile from a standing start. Using the *Slope Law*, plot the first derivative on the grid at the bottom of the drawing. Show all calculations.

Since the slope of the given curve is measured in units of kilometers per hour per second, the derivative curve shows the acceleration of the automobile at any instant (see Table 31.2). Add the scale designations and a caption for the ordinate axis of the derivative curve. Also add a suitable title in the space provided in the grid area. Match the style used for the given title and complete the grid lines. Fill in the blank for the acceleration at 10 seconds.

Drawing 95. Graphical Integration. References: §§31.12, 31.13. Plot the given data to describe the shape of one-fourth the cross-sectional area of a fuel tank. Integrate to determine the area. Show all calculations. Compute the volume of a 1200 mm long tank (see the figure). Show the equation and the result in the space provided. Add a suitable title in the space provided in the grid area. Match the style of lettering given and complete the grid lines.

Fuel Tank—One-Fourth Cross Section

Width, mm	*Height, mm*
0	300
100	296
200	283
300	260
400	224
500	166
550	120
600	0

If assigned, lay out a line EF (Empty–Full) along the right-hand margin to an appropriate scale representing the height of the tank, and calibrate the line at each 50 liters as a measuring stick (1 liter = 1000 cm^3).

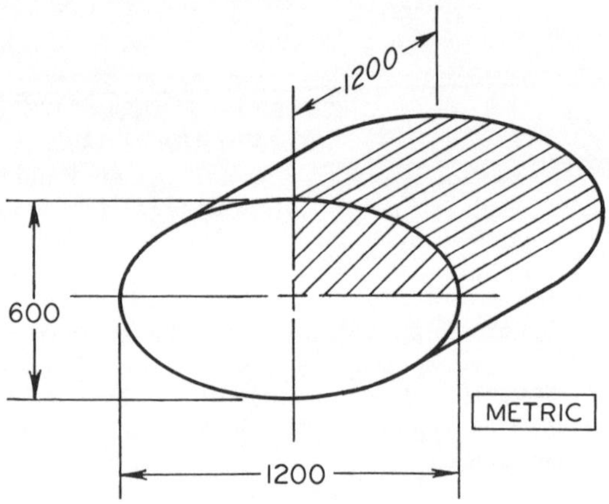
1200
600
METRIC
1200

FUEL TANK

Drawing 96. Computer-Aided Drafting—Terms and Descriptions. References: Chapters 3 and 8; Appendix 3. Some terms related to computer graphics are given in the table. A list of descriptions for these terms is given on the right. Find the matching description for each term and enter its letter identifier in the table.

Drawing 97. Computer-Aided Drafting—Two-Dimensional Coordinate Plot. Reference: Chapter 8.

Space 1. Digitize the single-view drawing by defining the X and Y coordinates of the indicated points and fill in the given table. Point A is the origin with values X and Y equal to zero. Consider each division of the grid as 1 unit. Keep in mind that any X values to the left of the origin and any Y values below the origin are negative.

Space 2. From the X and Y coordinate data given in the table, plot all points on the grid and complete the drawing. Point A is the origin. Consider each division of the grid as 1 unit.

Drawing 98. Computer-Aided Drafting—Three-Dimensional Coordinate Plot. Reference: Chapter 8. In drawing an image, the actions of the pen are Move and Draw.

Move: The pen moves from its present position to new X, Y, and Z coordinates specified. A line is not drawn. Numeral 0 is used to indicate Move action.

Draw: A line is drawn from the present pen position to new X, Y, and Z coordinates specified. Numeral 1 is used to indicate Draw action.

Space 1. Determine X, Y, and Z coordinates for all the points of the object. Complete the table for drawing the object, starting with point A. Coordinates X, Y, and Z are positioned as indicated by the arrows, with point A as origin. Try to use a minimum number of Move actions.

Space 2. According to the data shown in the table, draw the object on the grids provided. Coordinates X, Y, and Z are positioned as indicated by the arrows, with point A as origin.

Drawing 99. Computer-Aided Drafting—Menu Usage. Reference: Chapter 8. The drawing shows the front view of a Bracket that is to be generated on a graphics terminal. The numbers 1 through 21 refer to graphic entities that make up the drawing. Available menu commands for generating entities are given on the right. Complete the table by determining the menu commands to generate the entities. Enter the letter identifiers (A, B, etc.) of the menu selections in the table.

Drawing 100. Computer-Aided Drafting—Coordinate Systems. Reference: Chapter 8. Using the given descriptions for VIEW COORDINATES and WORLD COORDINATES, complete the tables for the front and right-side views of the object. Point number 1 is considered as the origin. Each grid division is equal to 1 unit.

Drawing 101. Detail Drawings. References: Chapters 6–10, 13–15, §16.8. Draw or sketch the necessary views of the object assigned. Select appropriate scale and sheet size. Dimension completely using metric or decimal-inch dimensions as assigned.

Alternate Assignment: Using a CAD system, produce a hard-copy multiview drawing of the problem assigned. Dimension completely.

Drawing 102. Detail Drawings. References: Chapters 6–10, 13–15, §16.8. Draw or sketch the necessary views of the object assigned. Select appropriate scale and sheet size. Dimension completely using metric or decimal-inch dimensions as assigned.

Alternate Assignment: Using a CAD system, produce a hard-copy multiview drawing of the problem assigned. Dimension completely.

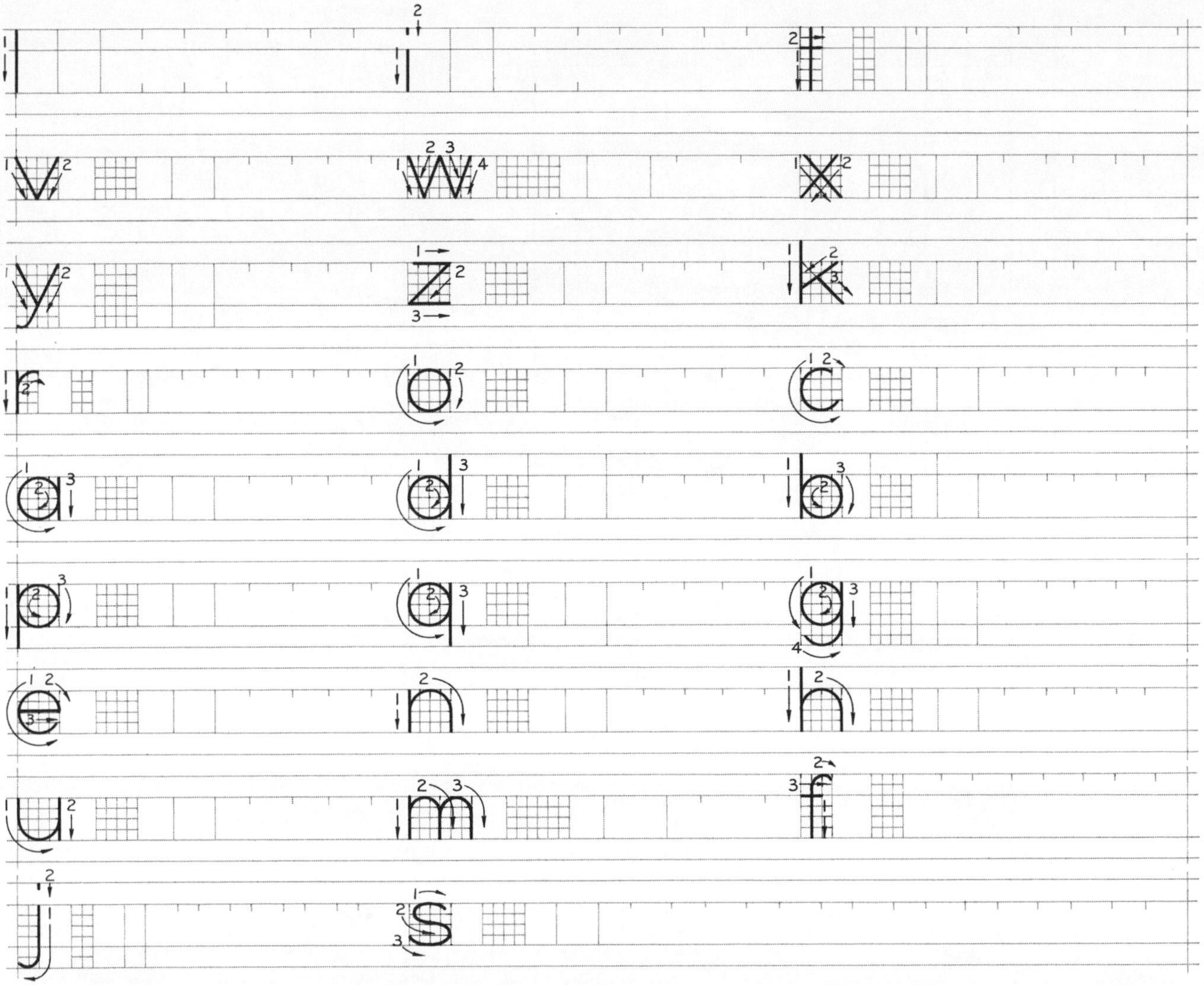

Now they are universally used for engineering

working drawings. They should not be executed

mechanically but should be made entirely freehand. The term "single-

stroke" does not imply that the letter is produced with a continuous

movement of the pencil or pen, but that the letter is made of one

LOWER CASE VERTICAL LETTERING	DRAWN BY	FILE NO.	DRAWING 1A

	DRAWN BY	FILE NO.	DRAWIN

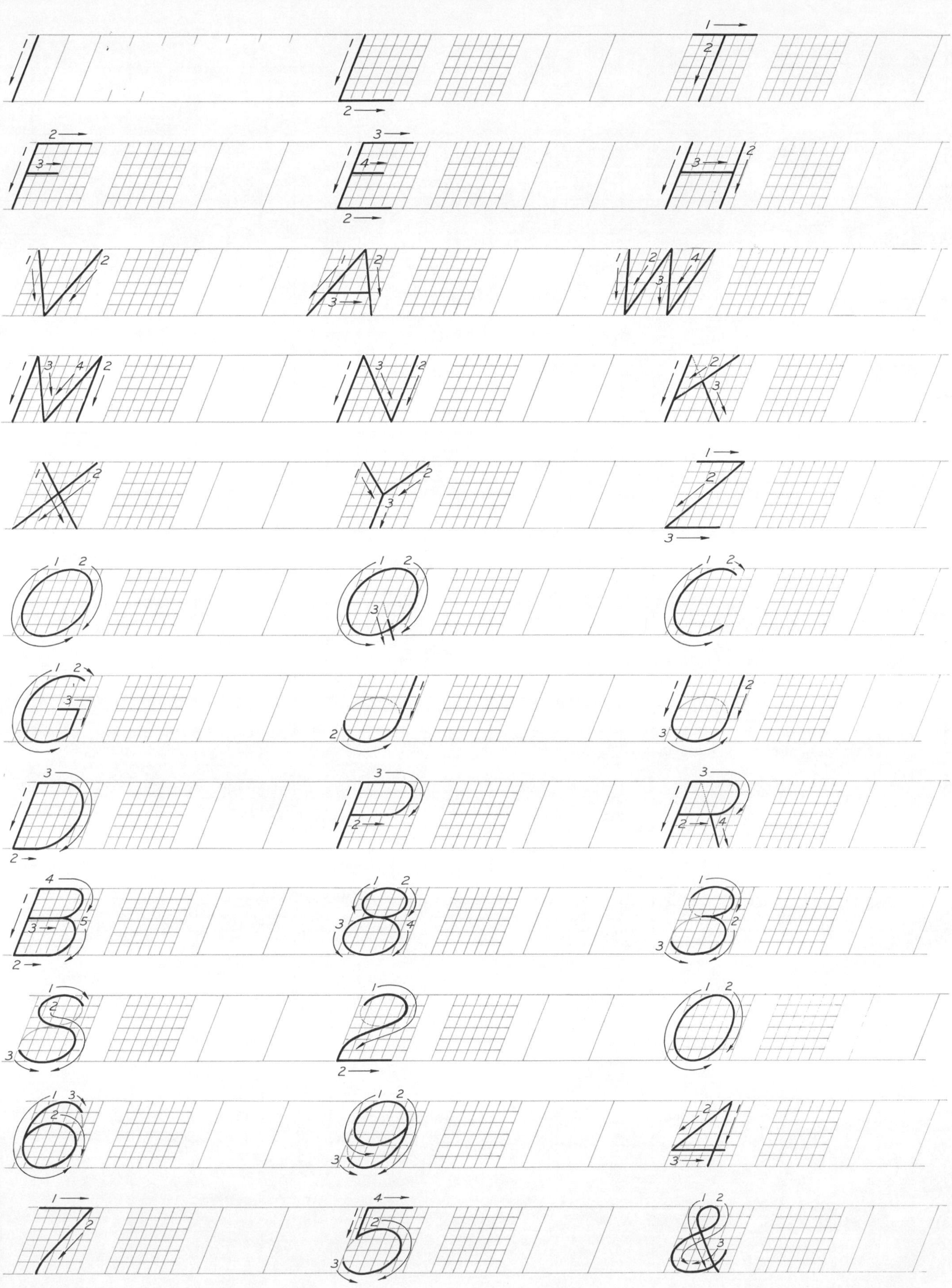

CAPITALS AND NUMERALS INCLINED LETTERING	DRAWN BY	FILE NO.	DRAWING 4

	DRAWN BY	FILE NO.	DRAWING

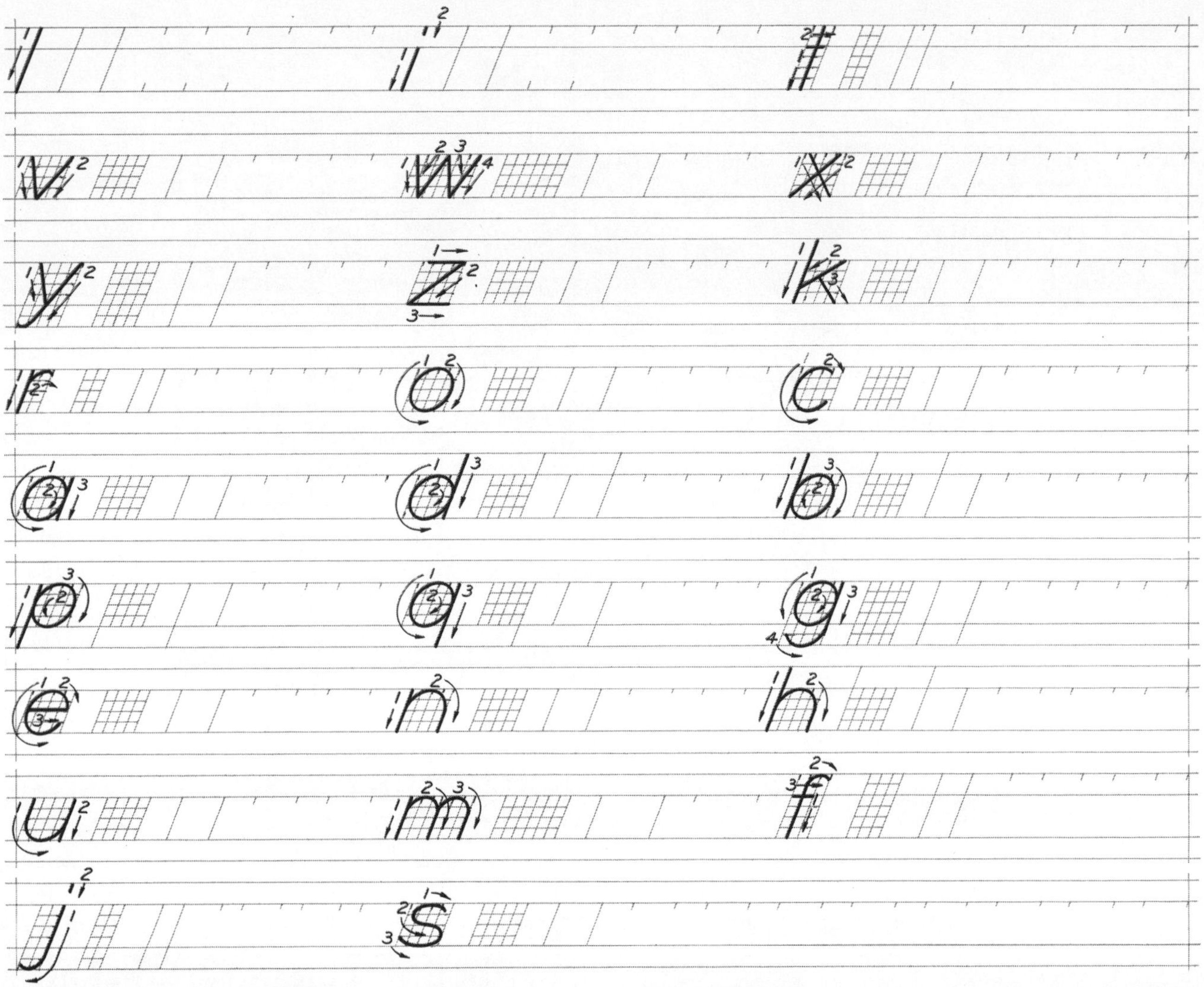

smaller letters. The lower-case inclined letters

may be regarded, like the upper-case inclined

letters, as oblique projections of vertical letters, in which all of

the circles in the vertical alphabet become ellipses in the in-

clined alphabet. As in inclined capital letters, all ellipses have

LOWER CASE INCLINED LETTERING	DRAWN BY	FILE NO.	DRAWING 4A

	DRAWN BY	FILE NO.	DRAWIN

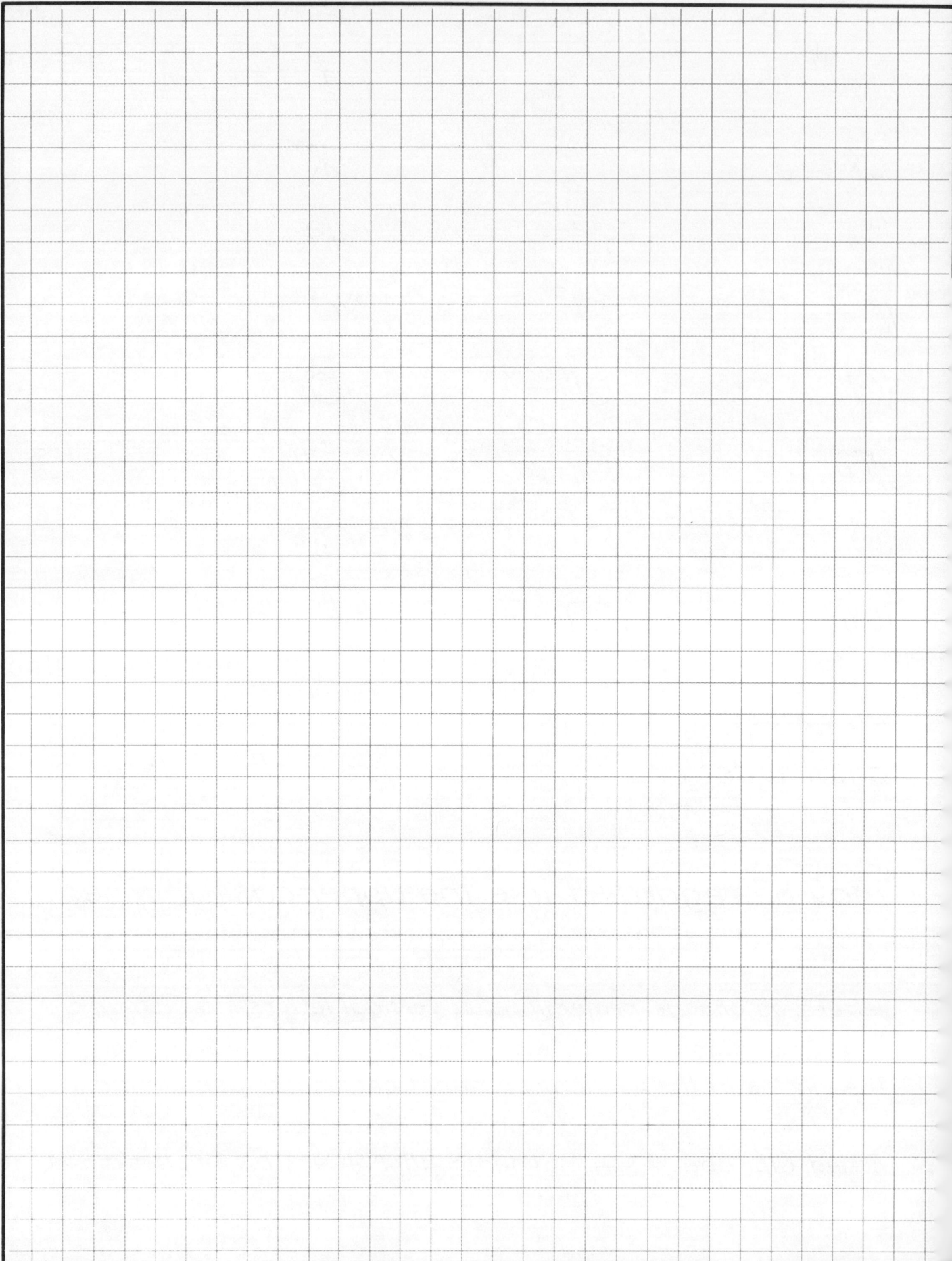

WHILE IT IS TRUE THAT

"PRACTICE MAKES PERFECT," IT

MUST BE UNDERSTOOD THAT

PRACTICE IS NOT ENOUGH, BUT IT

MUST BE ACCOMPANIED BY A CON-

TINUOUS EFFORT TO IMPROVE. EXCEL-

LENT LETTERERS ARE OFTEN NOT GOOD

WRITERS. USE A FAIRLY SOFT PENCIL, AND AL-

WAYS KEEP IT SHARP, ESPECIALLY FOR SMALL

LETTERS. MAKE THE LETTERS CLEAN-CUT AND

DARK—NEVER FUZZY, GRAY, OR INDEFINITE. 1234

$1\frac{1}{2}$ 1.500 $\frac{3}{16}$ 45'-6 32° 15.489 $\frac{13}{64}$ 12"=1'-0 $7\frac{5}{16}$ 12.3 $\frac{1}{2}$ $2\frac{1}{4}$

ONE MUST HAVE A CLEAR MENTAL IMAGE OF THE LETTERS. 234

CAPITALS AND NUMERALS INCLINED LETTERING	DRAWN BY	FILE NO.	DRAWING 5

	DRAWN BY	FILE NO.	DRAWIN

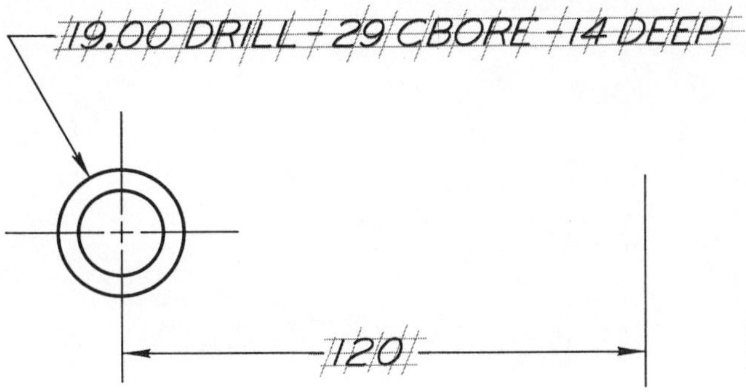
19.00 DRILL - 29 CBORE - 14 DEEP
120

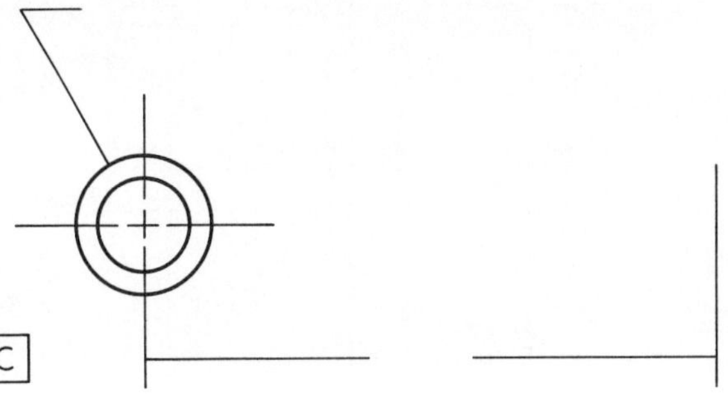

METRIC

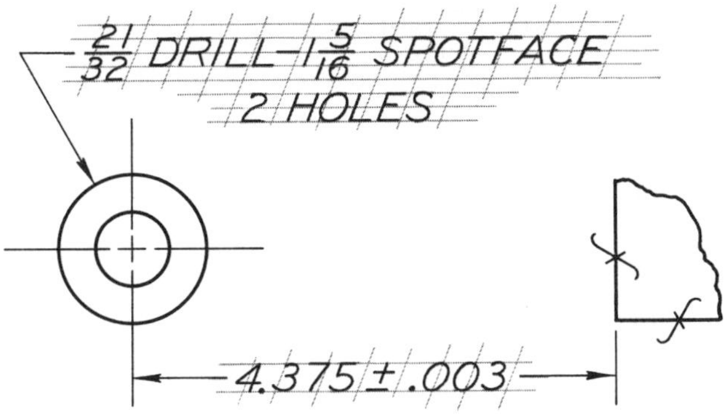
21/32 DRILL - 1 5/16 SPOTFACE
2 HOLES
4.375 ± .003

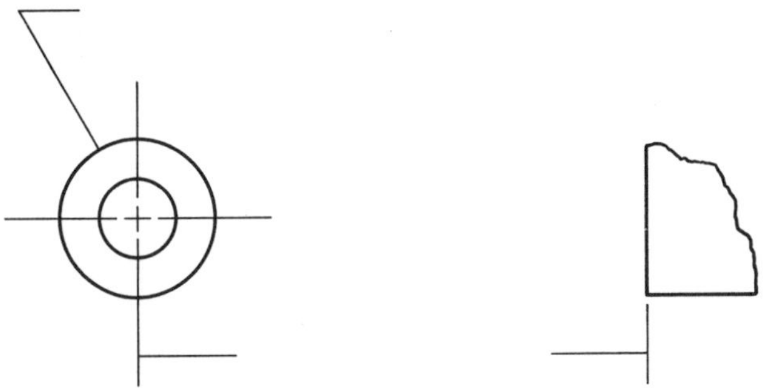

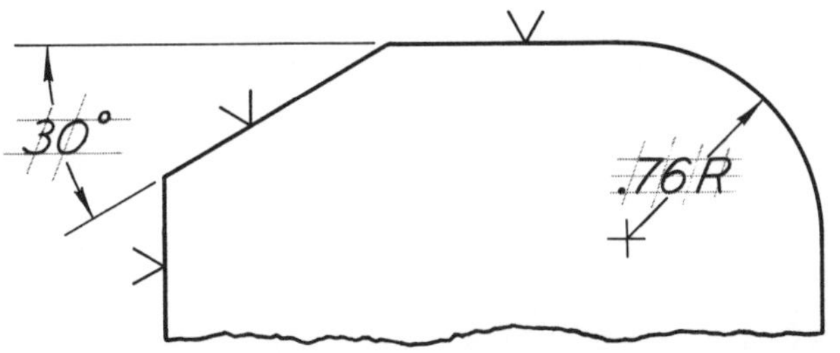
30°
.76R

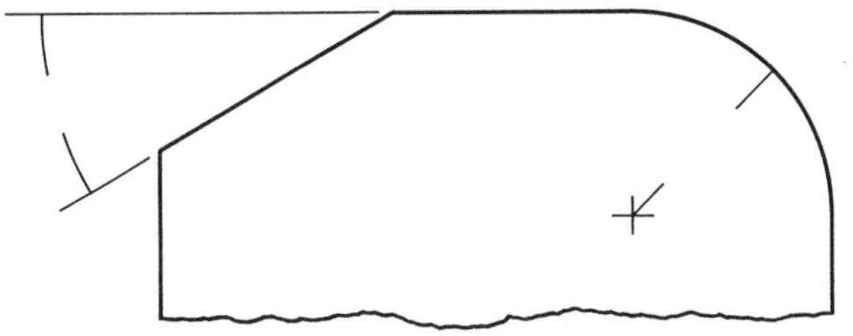

.06 x 45° CHAMFER BOTH ENDS

FILE FINISH AND POLISH

.562-.564 REAM-2 HOLES

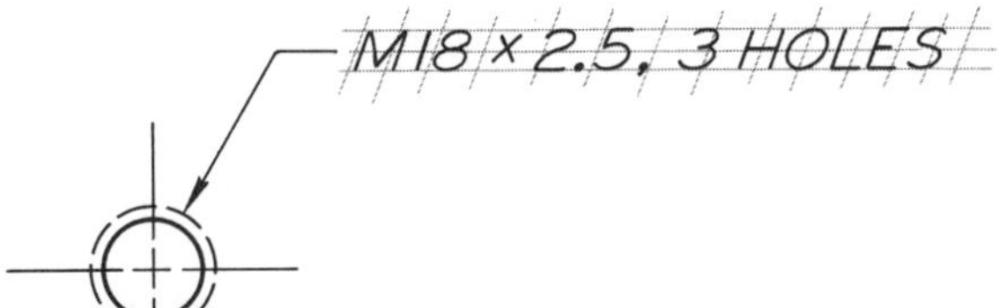
M18 x 2.5, 3 HOLES

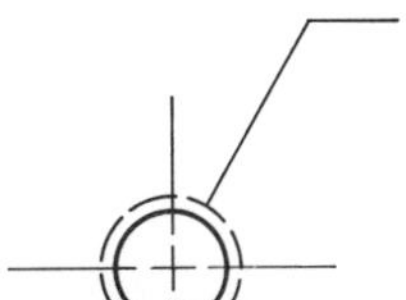

4 mm (5/32") *Tool Holder*

4 mm (5/32")

3 mm (1/8") *F A O - Cyanide & Polish*

2.5 mm (3/32")

3 mm (1/8") *M S - 3 Reqd*

SHOP NOTES AND DIMENSIONS	DRAWN BY	FILE NO.	DRAWING
INCLINED LETTERING			6

	DRAWN BY	FILE NO.	DRAWIN

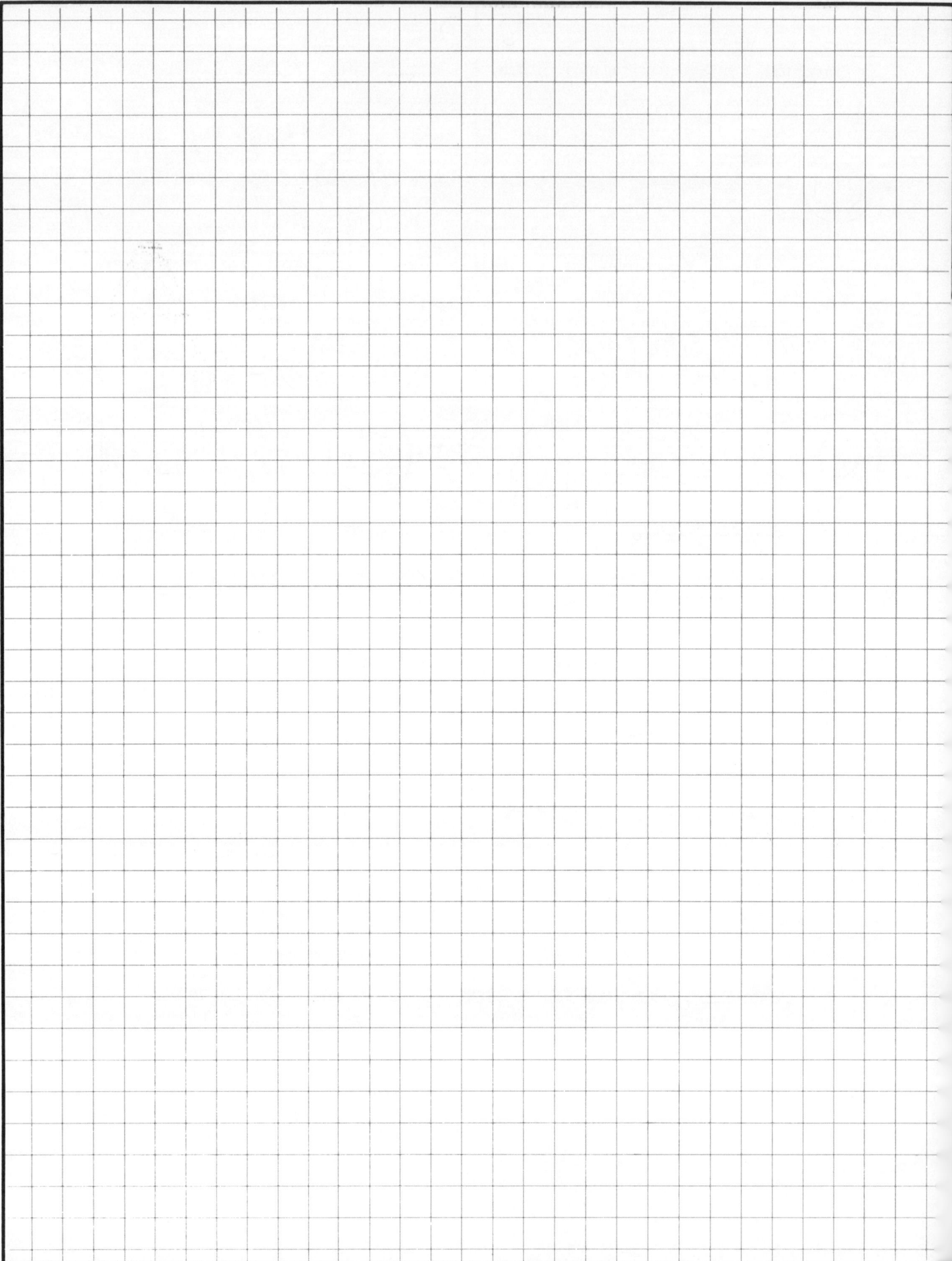

(1) Complete the view of the Lever. Construct all points of tangency.

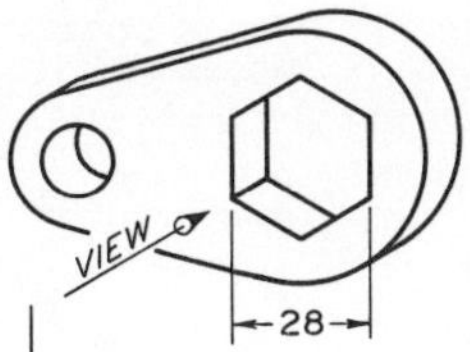
VIEW
28

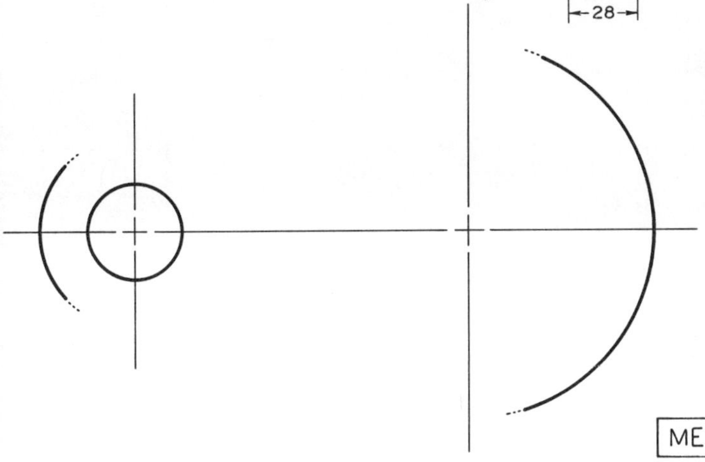

(2) Complete the view of the Clamp. Construct all points of tangency.

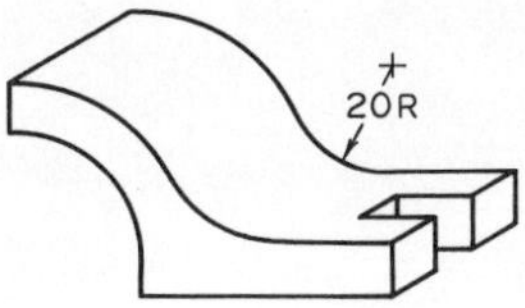
20R

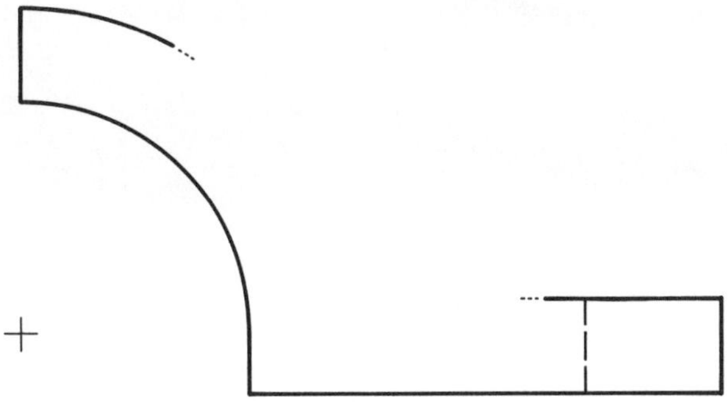

METRIC

(3) Complete the view of the Bracket. Construct all points of tangency.

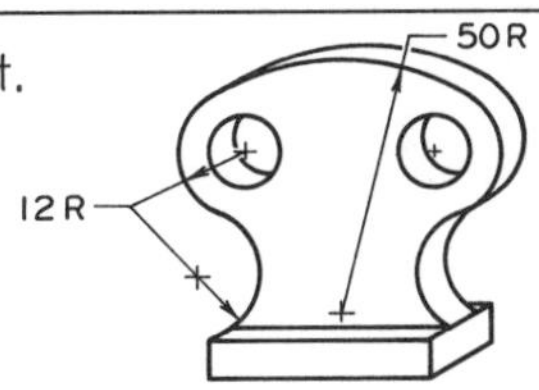
50R
12R

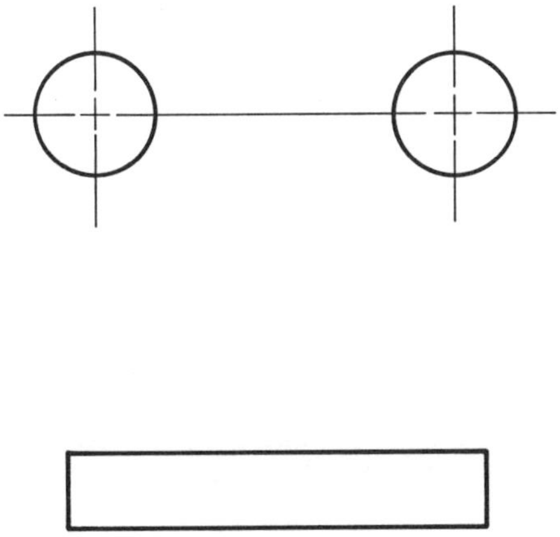

(4) Complete the view of the Link. Construct all points of tangency.

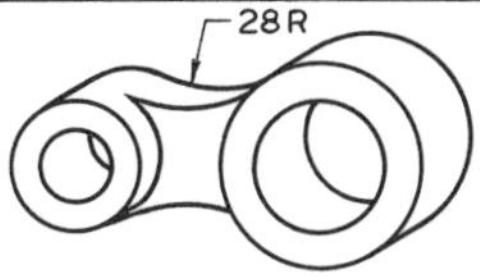
28R

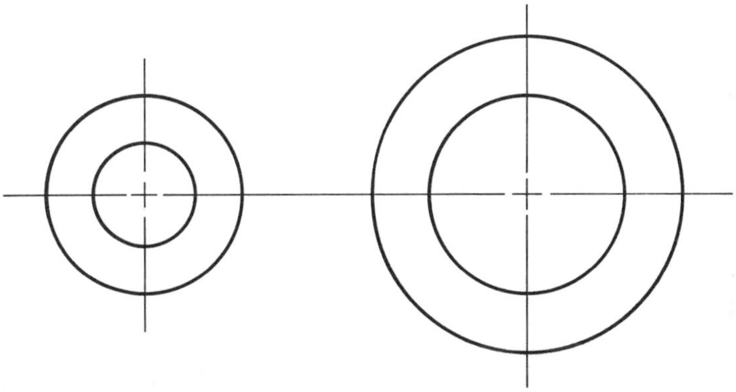

METRIC

(5) Complete the view of the Toggle Link. Construct all points of tangency.

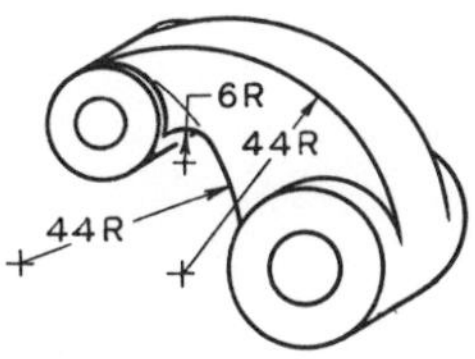
6R
44R
44R

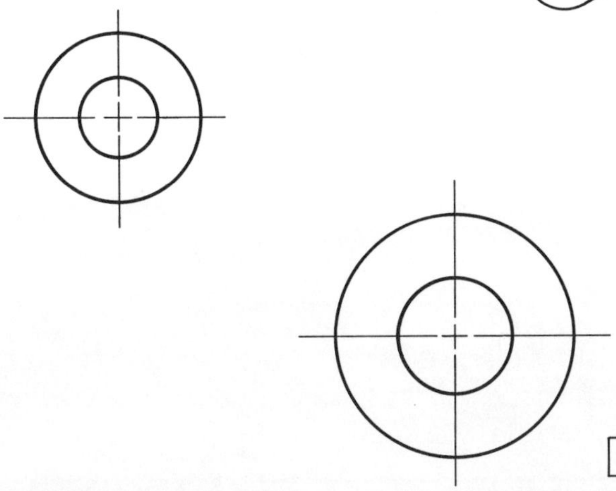

(6) Complete the view of the Bearing using approx. 4-center ellipse. Construct all points of tangency.

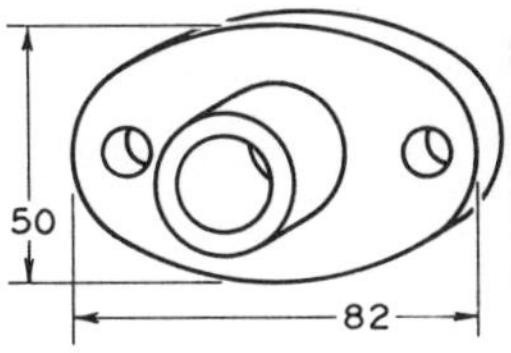
50
82

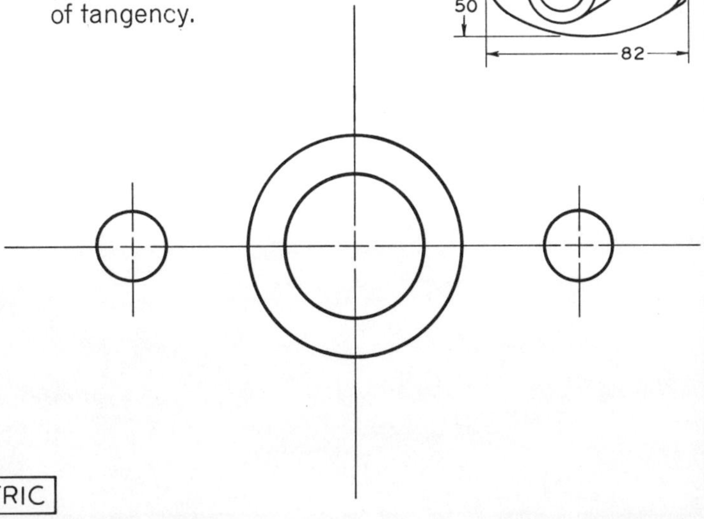

METRIC

TANGENCIES GEOMETRIC CONSTRUCTIONS	DRAWN BY	FILE NO.	DRAWING 11

	DRAWN BY	FILE NO.	DRAWIN

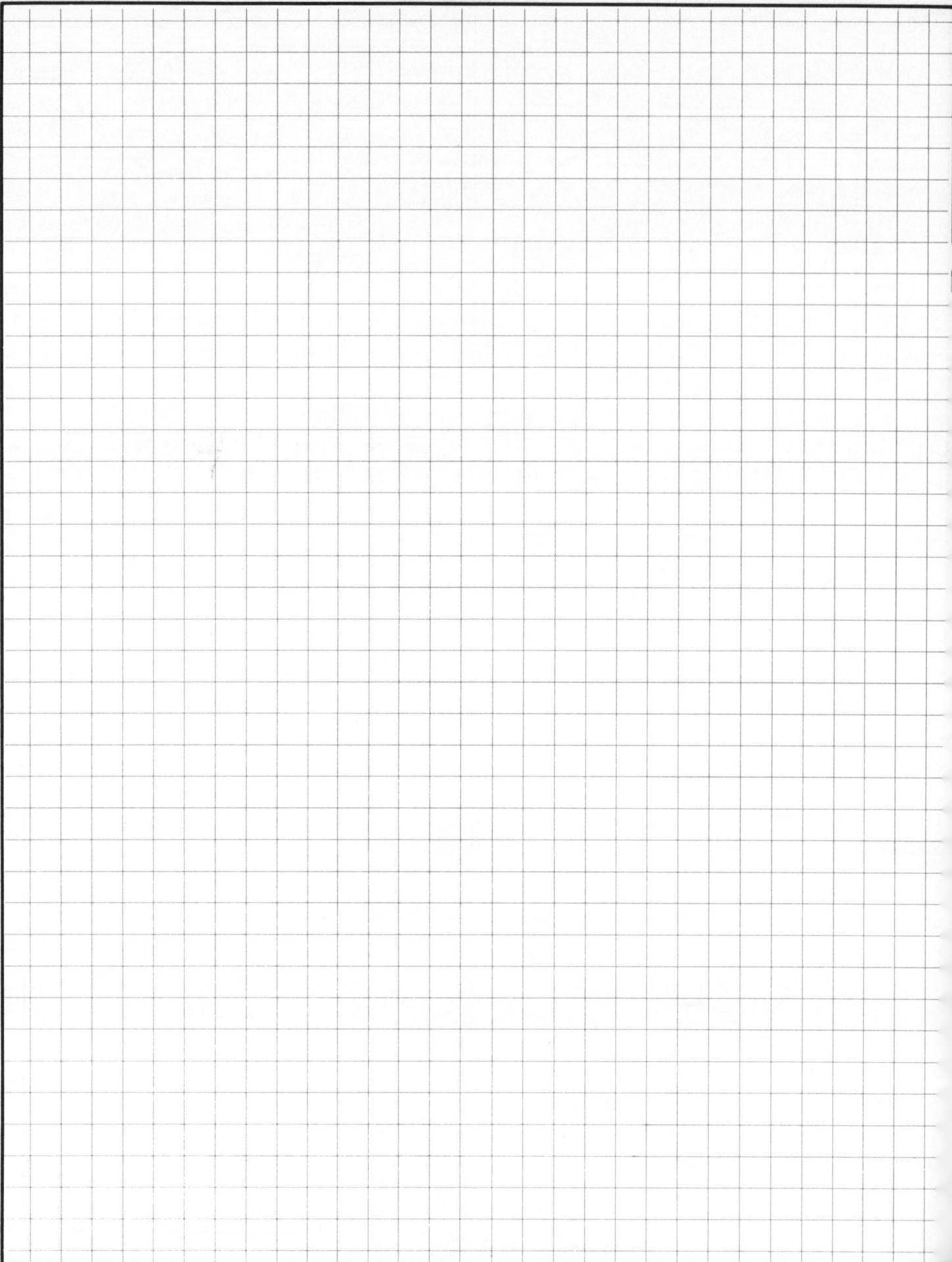

(1) Complete the view of the elliptical Cam.

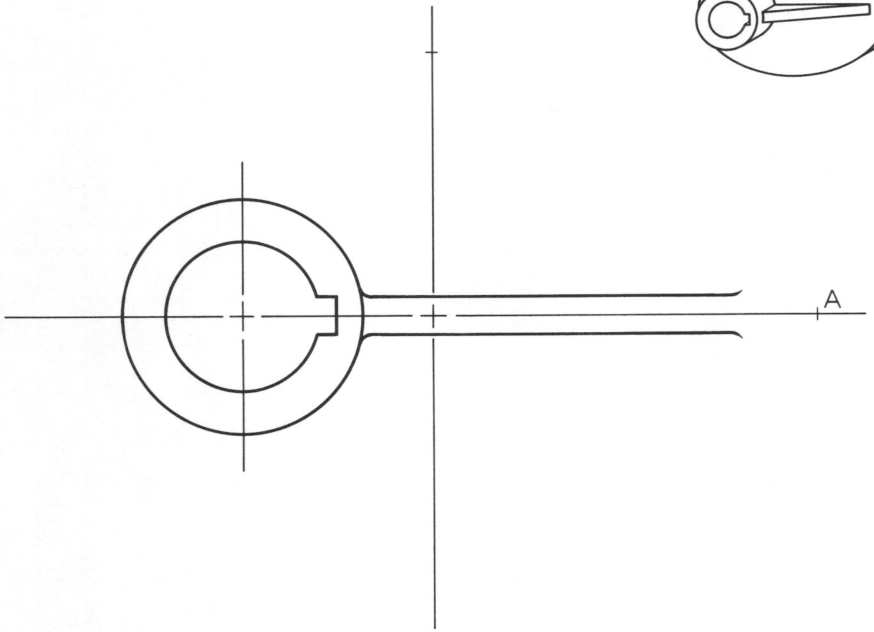
A

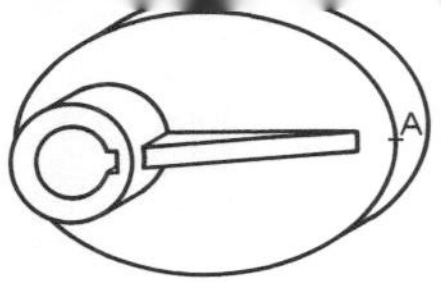
A

(2) Design a parabolic arch with a span of 102 m along AB and a rise of 64 m.
Scale: 1/1000

A
B

(3) Design a parabolic curve between XO and YO and complete the view of the Sliding Shoe.

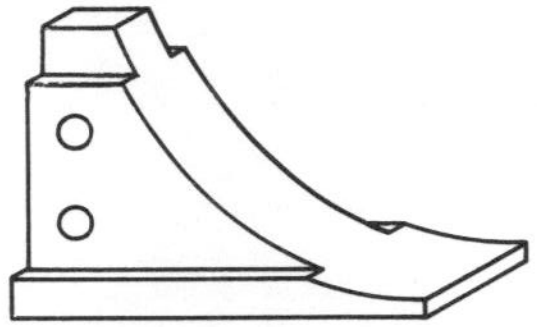

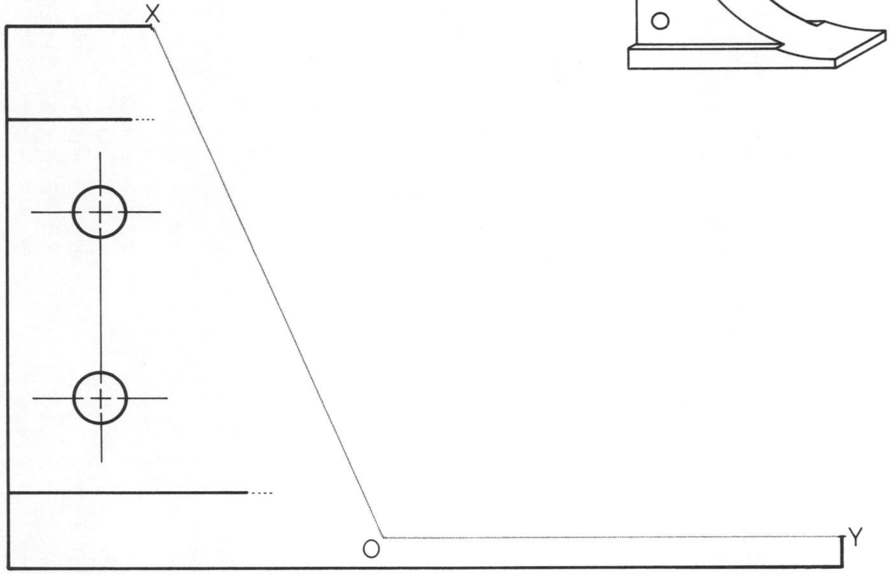
X
O
Y

(4) Draw a right-hand helix with a lead of 60 mm. Start helix at point A. Show visibility.

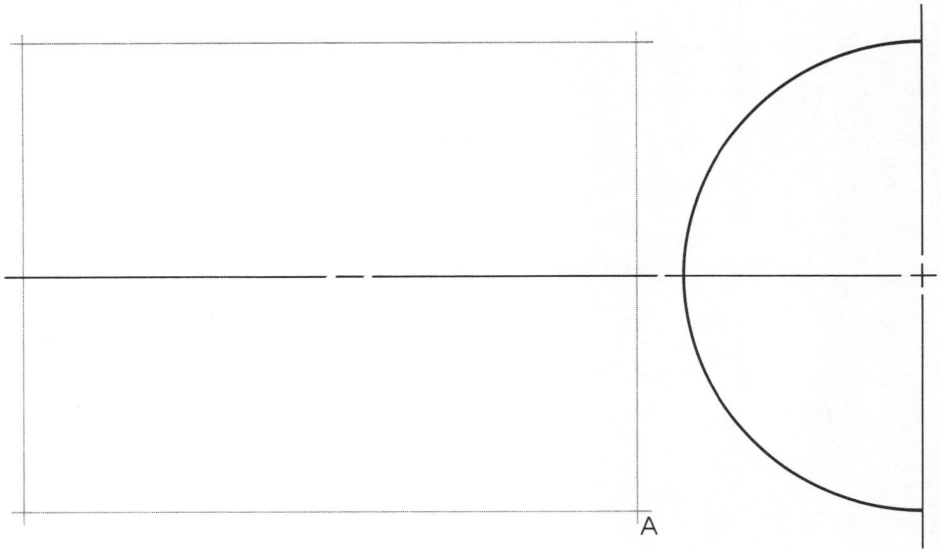
A

ELLIPSE, PARABOLA, AND HELIX

GEOMETRIC CONSTRUCTIONS

DRAWN BY	FILE NO.	DRAWING
		12

	DRAWN BY	FILE NO.	DRAWING

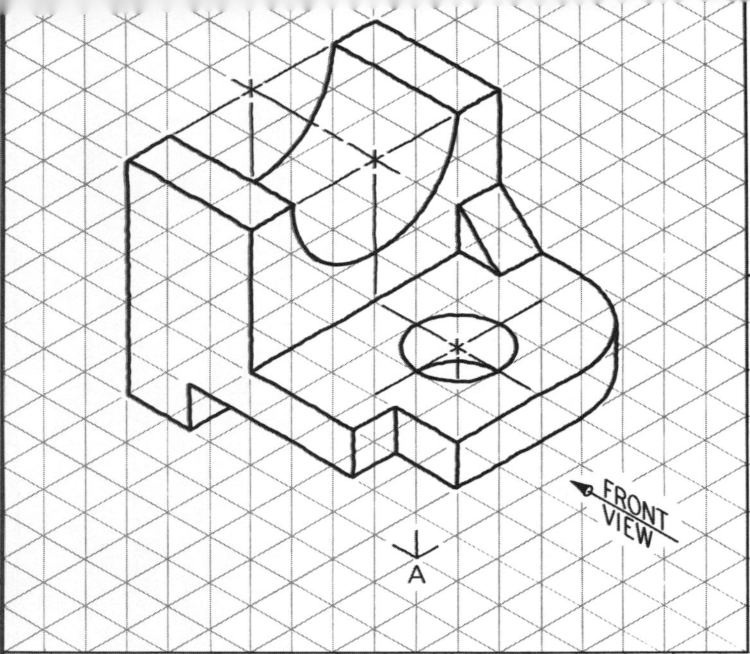

FRONT VIEW
A

SPECIAL BEARING

Make a full-size 6-view freehand sketch of the part, with the views in the standard arrangement.

Label all views: FRONT, TOP, etc. Space the views two squares apart. Show all hidden lines and center lines.

A

FRONT

The necessary views for a complete shape description are:

MULTIVIEW
TECHNICAL SKETCHING

DRAWN BY

FILE NO.

DRAWING 13

	DRAWN BY	FILE NO.	DRAWIN

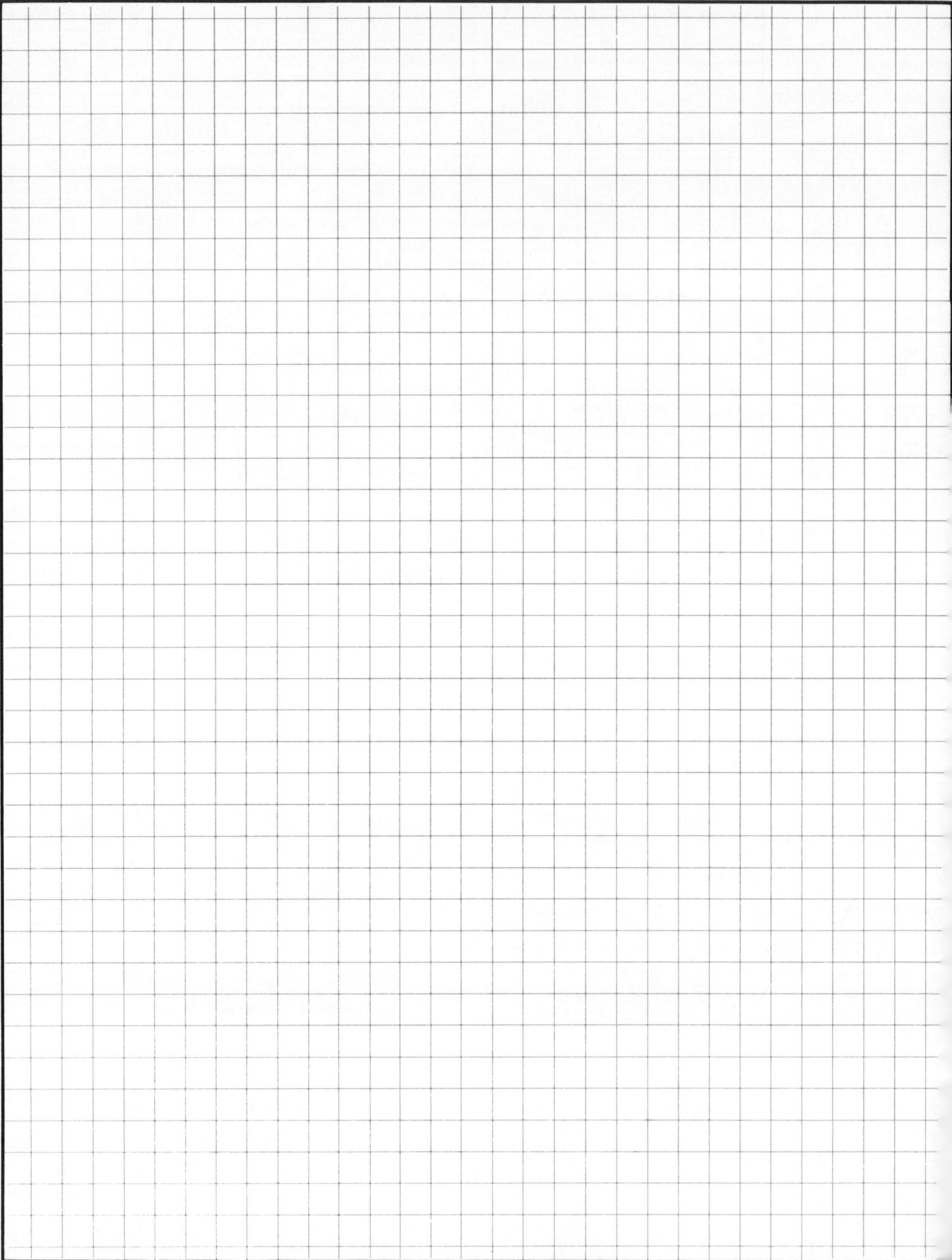

1 ROD BASE
Sketch front, top and right-side views.

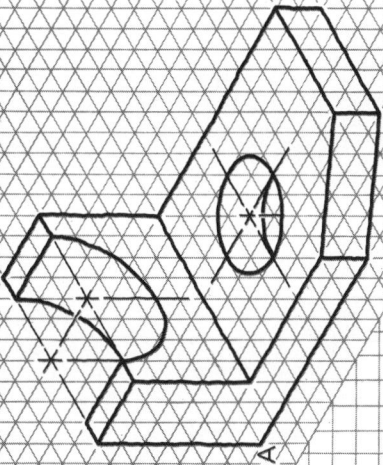

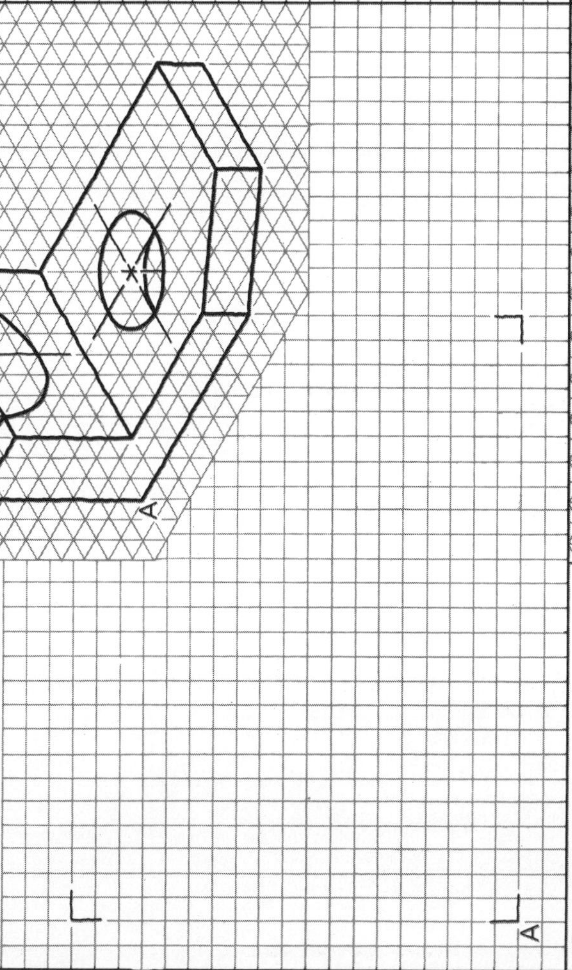

2 STOP SUPPORT
Sketch top view and isometric.

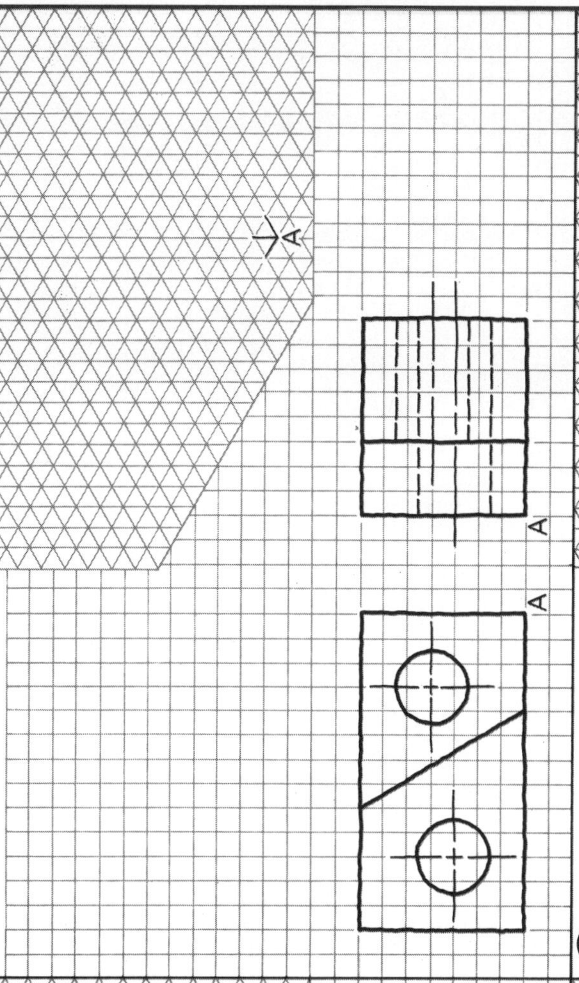

3 ADJUSTABLE GUIDE
Sketch top view and complete isometric.

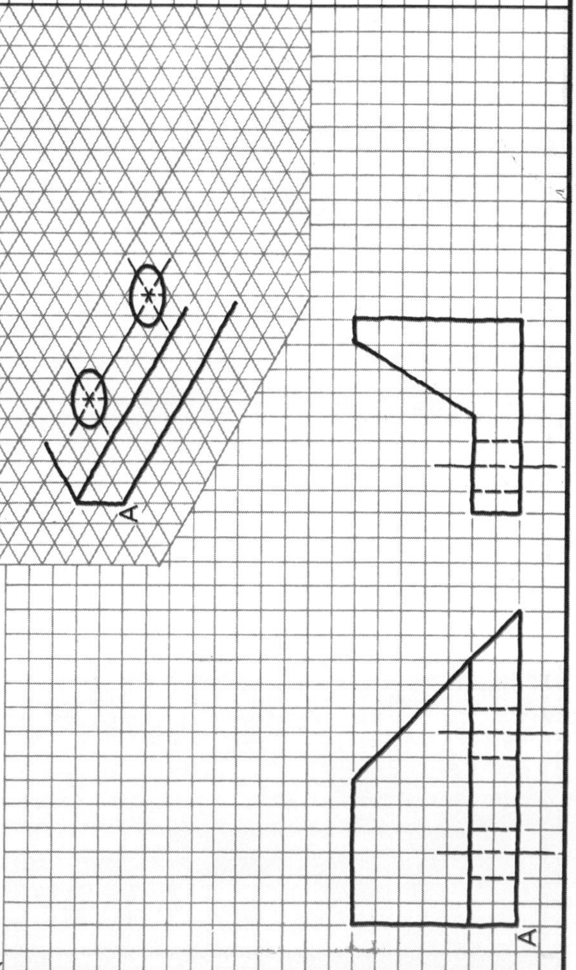

4 WEDGE
Sketch isometric and complete top view.

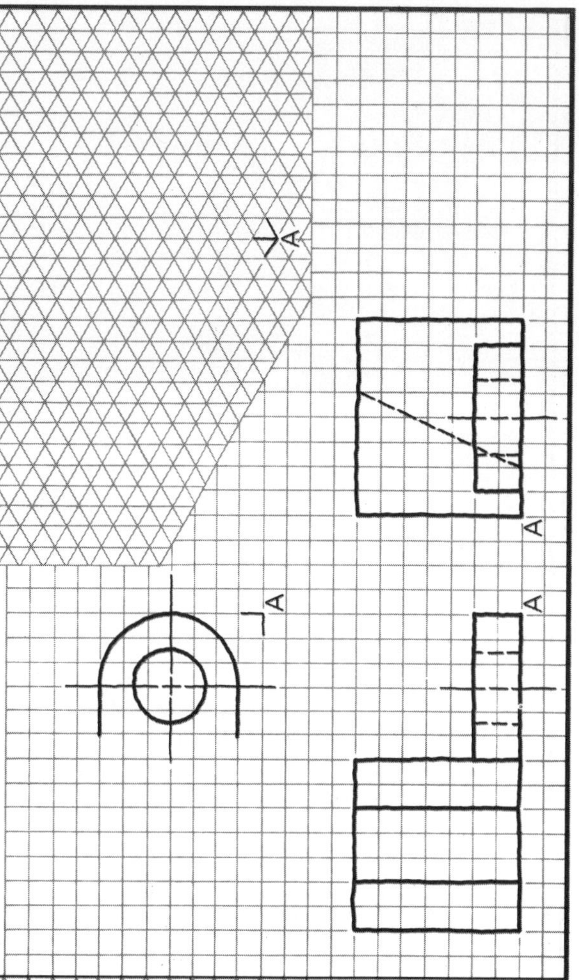

DRAWN BY

MULTIVIEW AND ISOMETRIC TECHNICAL SKETCHING	DRAWN BY	FILE NO.	DRAWING 14

	DRAWN BY	FILE NO.	DRAWI

(1)

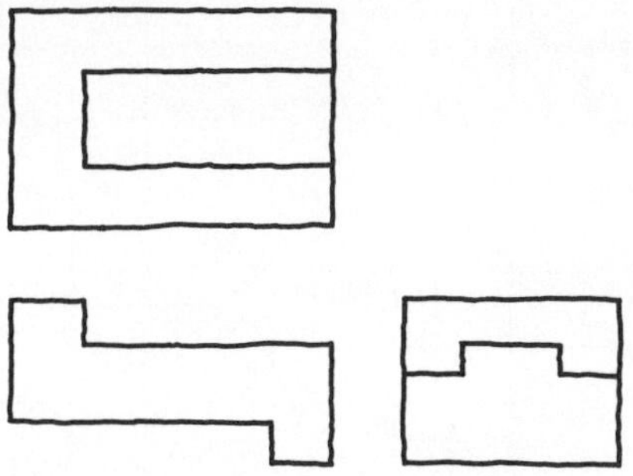

(2)

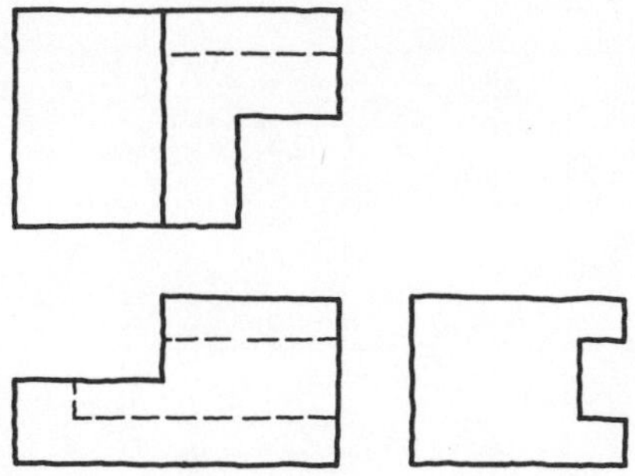

(3)

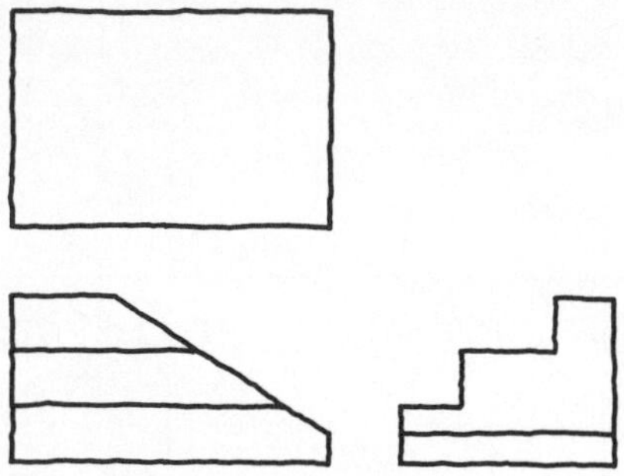

(4)

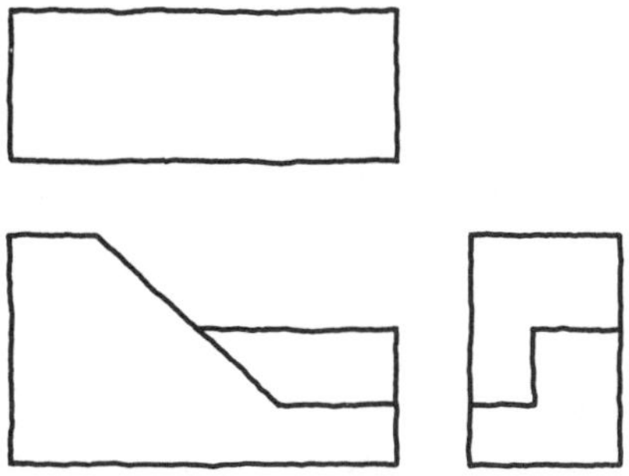

(5)

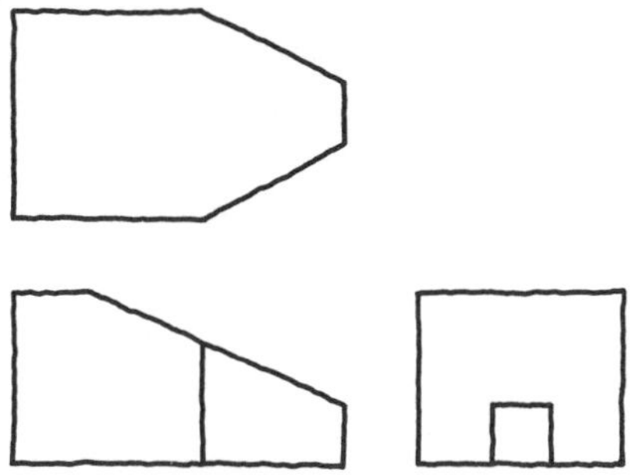

(6)

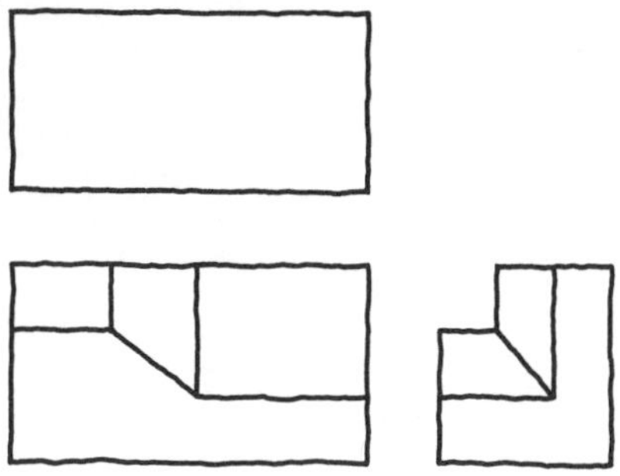

(7)

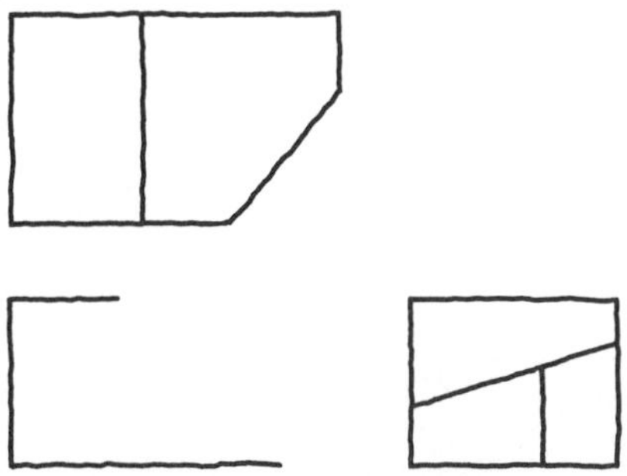

(8)

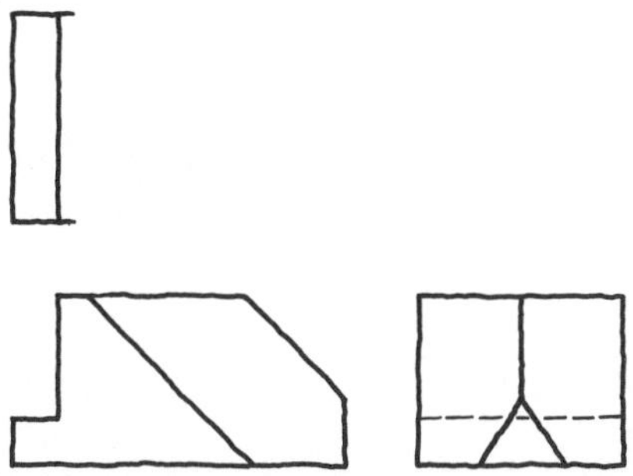

(9)

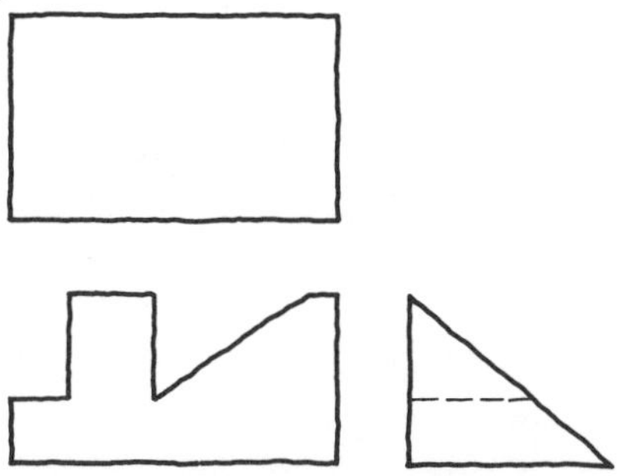

(10)

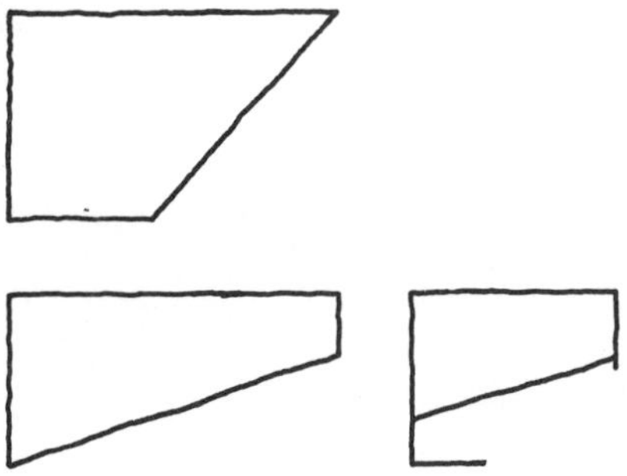

(11)

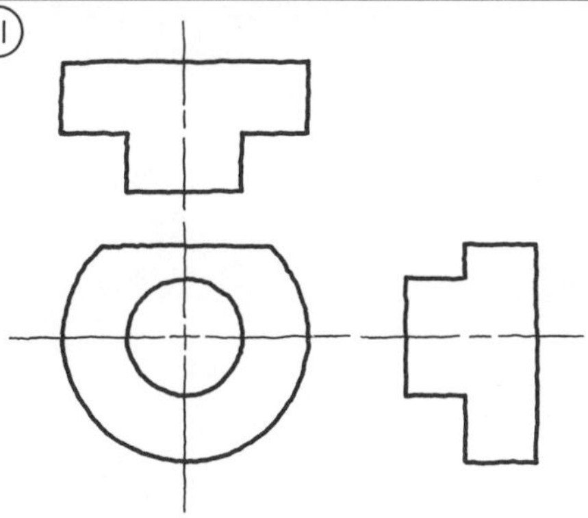

(12)

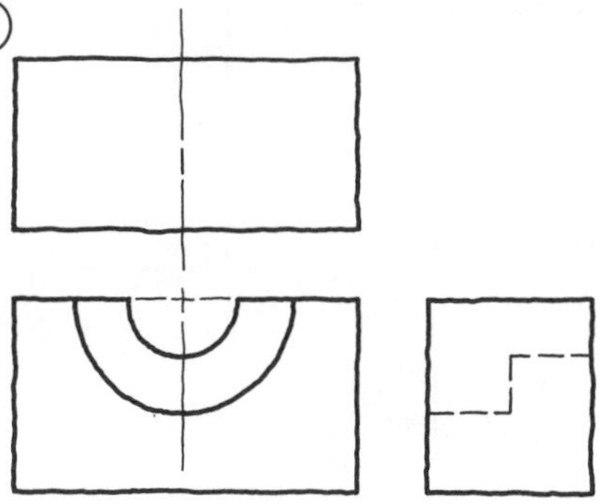

(13)

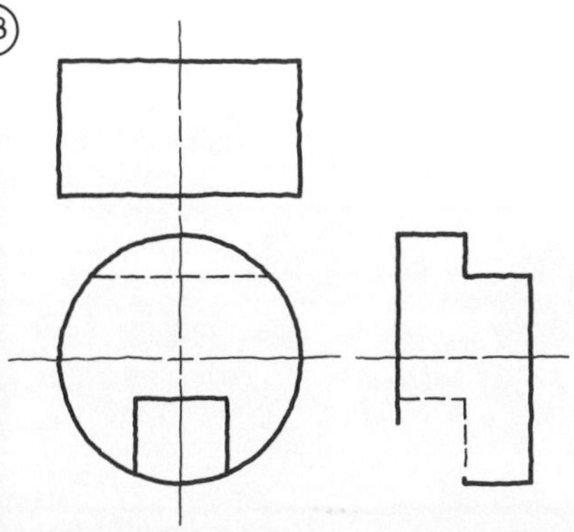

(14)

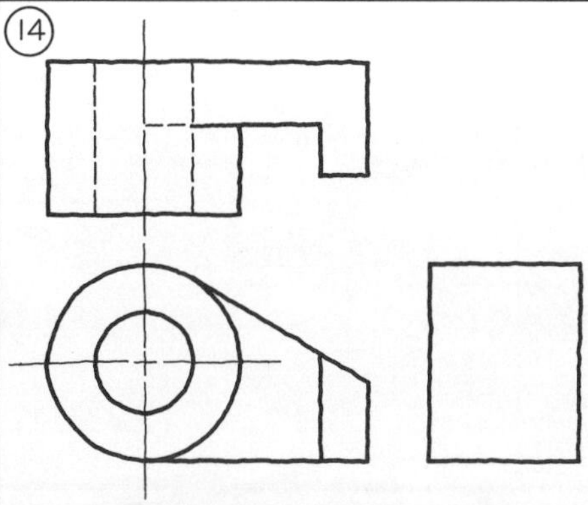

(15)

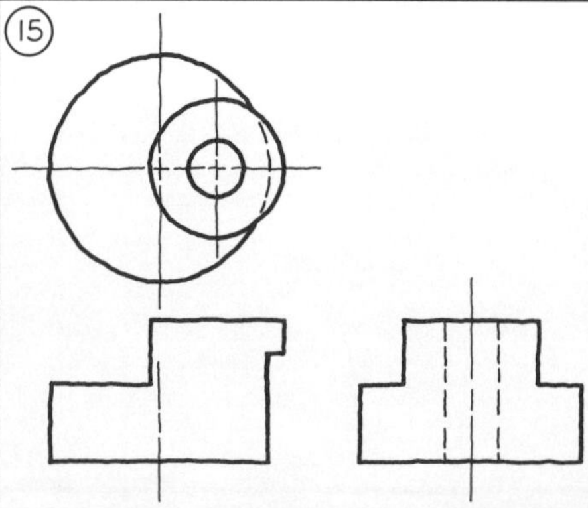

MISSING LINES MULTIVIEW PROJECTION	DRAWN BY	FILE NO.	DRAWING 15

	DRAWN BY	FILE NO.	DRAWIN

HOLDER PUSH DOG

Sketch front, top, and left-side views.

A

FREEHAND

MULTIVIEW PROJECTION

DRAWN BY

FILE NO.

DRAWING 16

	DRAWN BY	FILE NO.	DRAWI

① STOP BLOCK

Sketch top view.

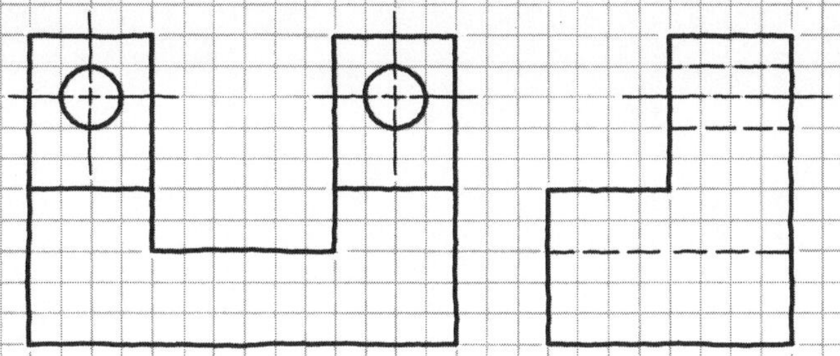

② LOCK SLIDE

Sketch right-side view.

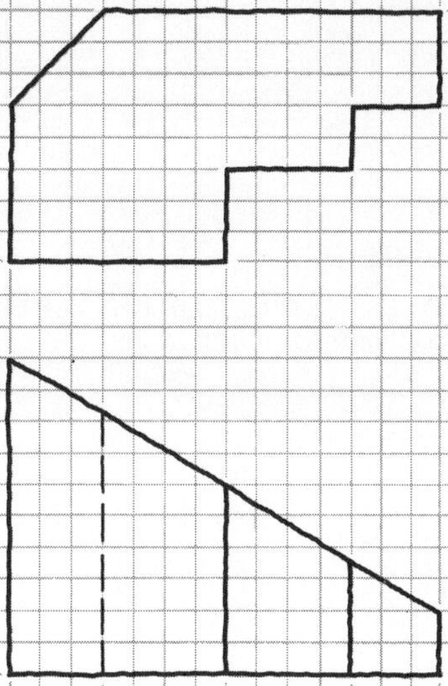

③ SUPPORT

Sketch top view.

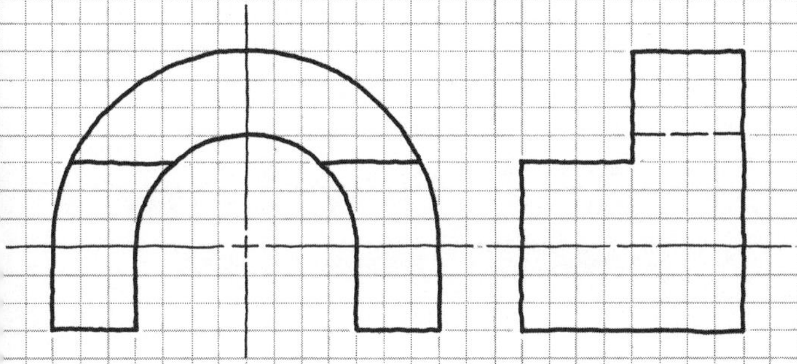

④ WEDGE

Sketch right-side view.

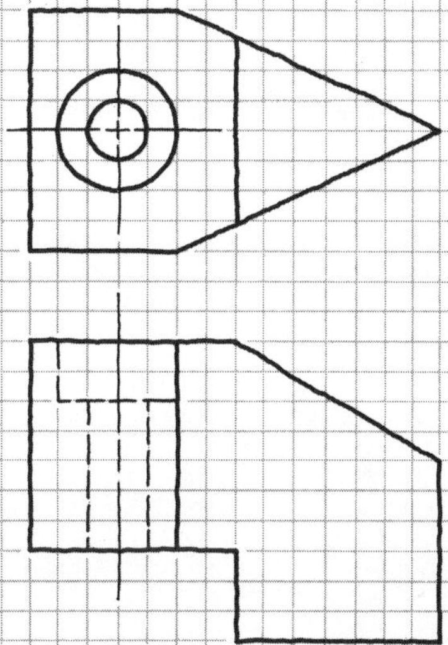

⑤ KEY

Sketch right-side view.

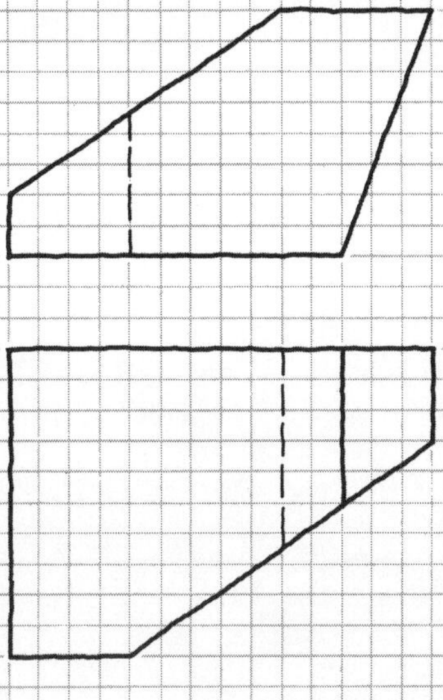

⑥ GUIDE

Sketch front view.

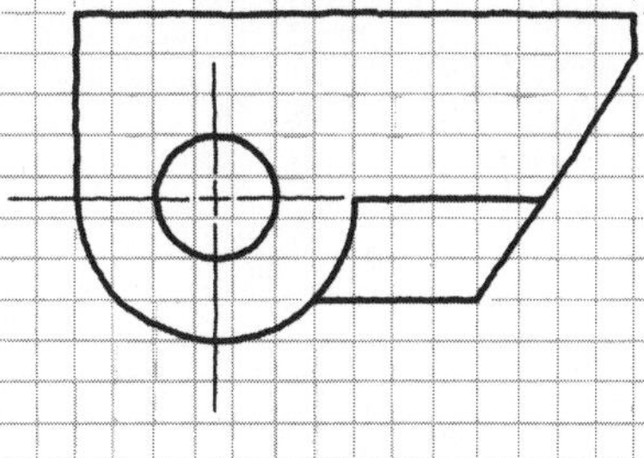

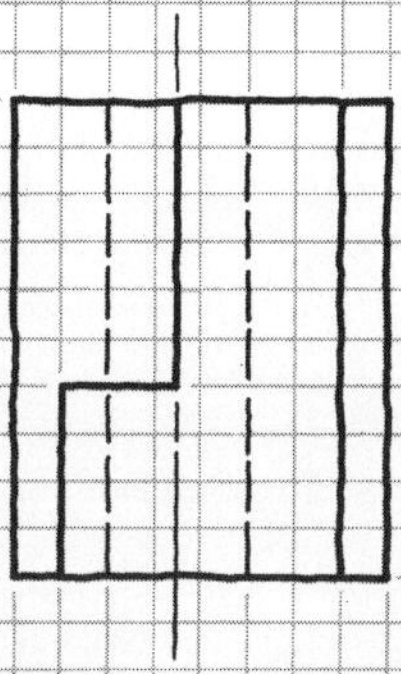

MISSING VIEWS MULTIVIEW PROJECTION	DRAWN BY	FILE NO.	DRAWING 17

	DRAWN BY	FILE NO.	DRAW

1

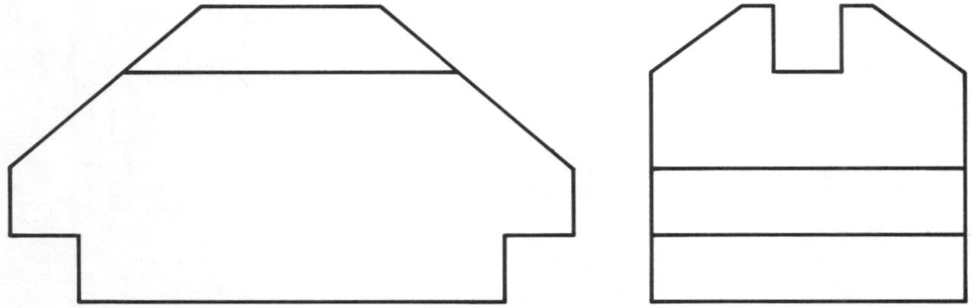

GAGE BASE
Draw top view.

2

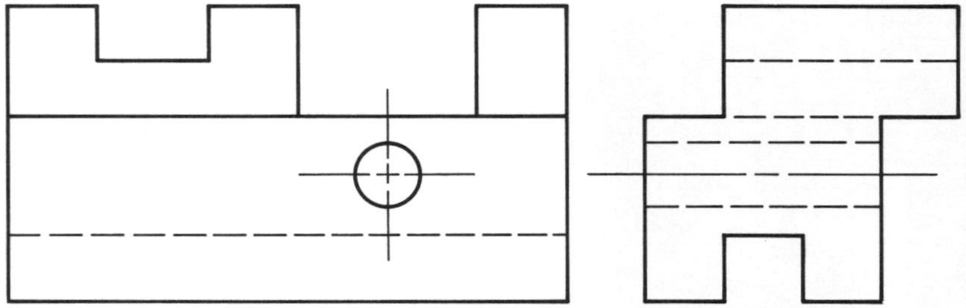

STOP BLOCK
Draw top view.

3

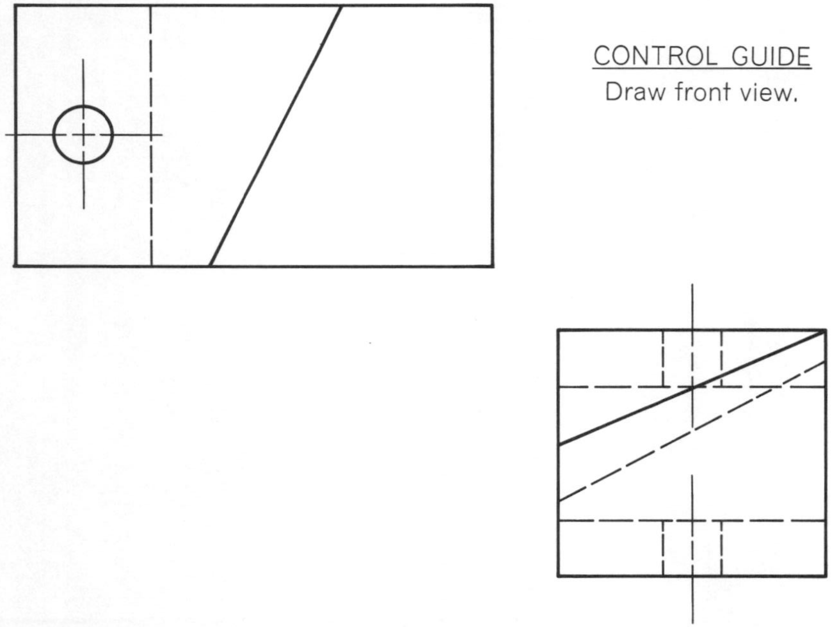

CONTROL GUIDE
Draw front view.

4

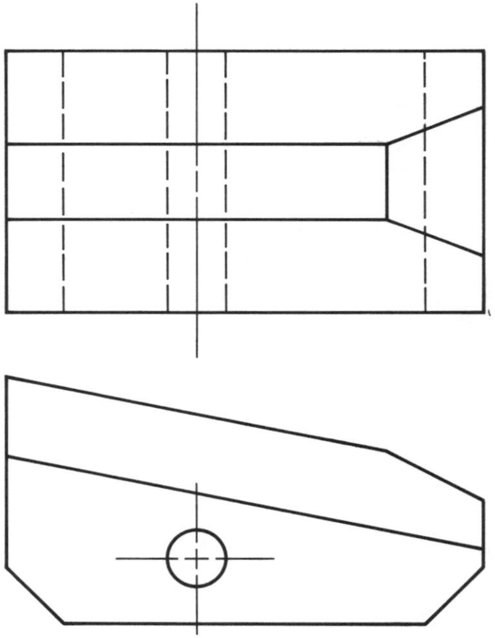

WEDGE
Draw side view.

MISSING VIEWS
MULTIVIEW PROJECTION

DRAWN BY	FILE NO.	DRAWING
		18

	DRAWN BY	FILE NO.	DRAWIN

(1)

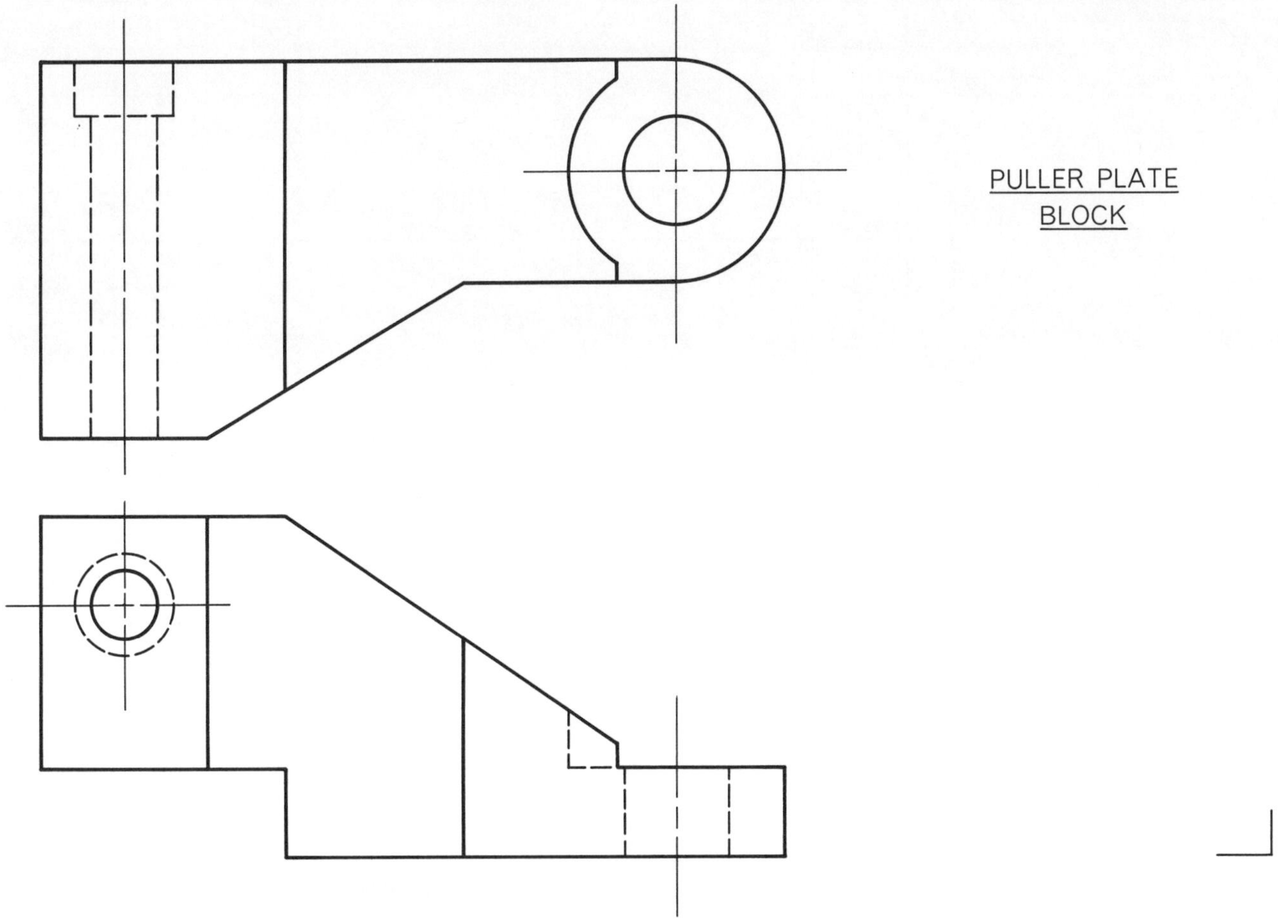
PULLER PLATE
BLOCK

(2)

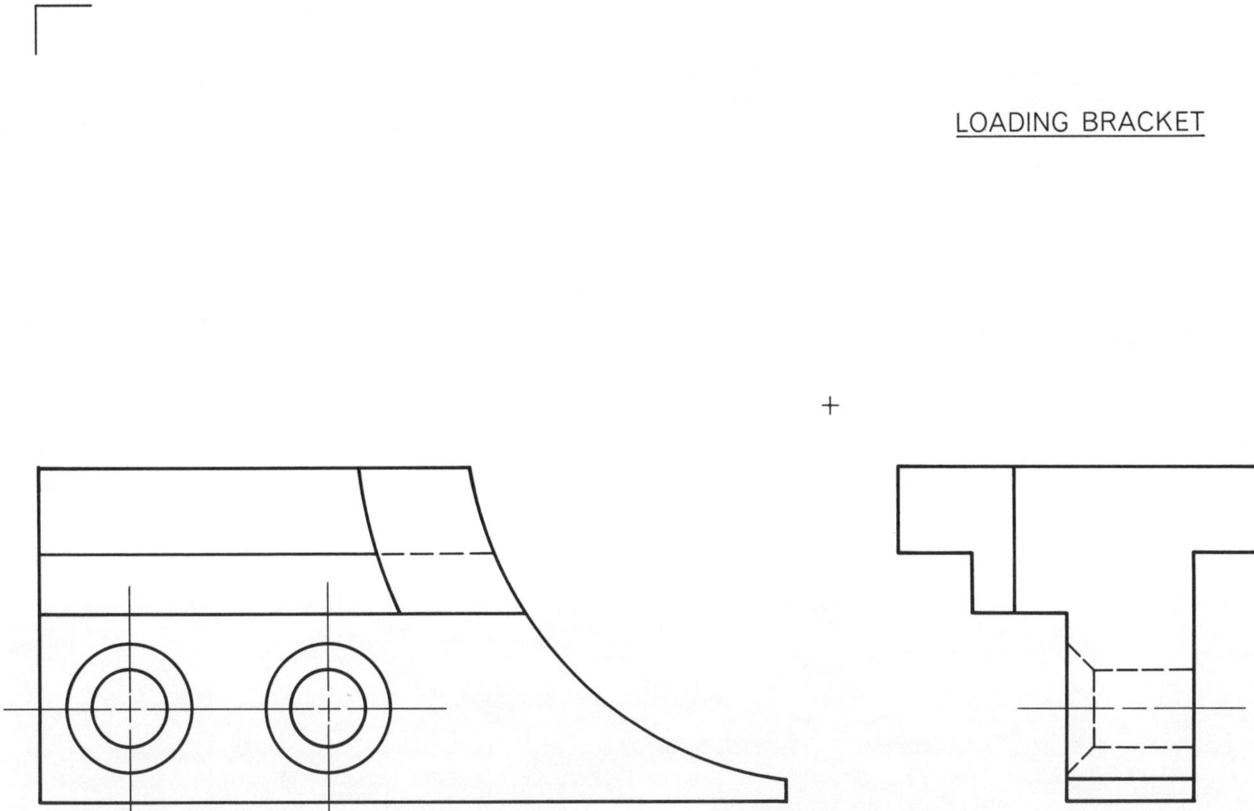
LOADING BRACKET

MISSING VIEWS

MULTIVIEW PROJECTION

DRAWN BY

FILE NO.

DRAWING 20

	DRAWN BY	FILE NO.	DRAWI

①

BORING TOOL

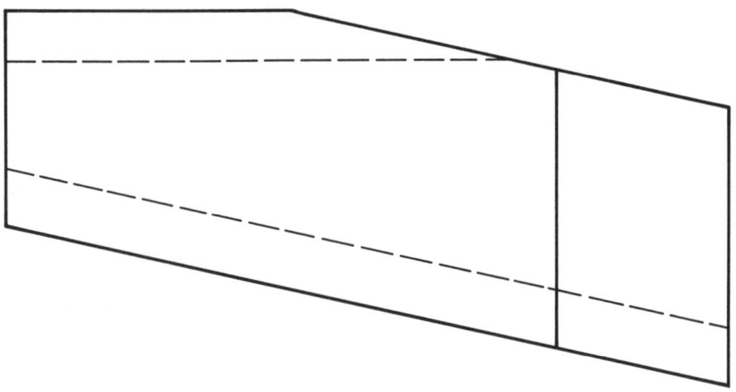

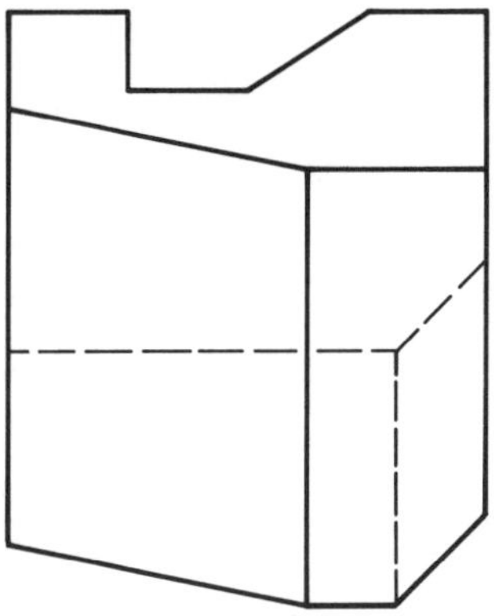

②

THREAD ATTACHMENT CAM

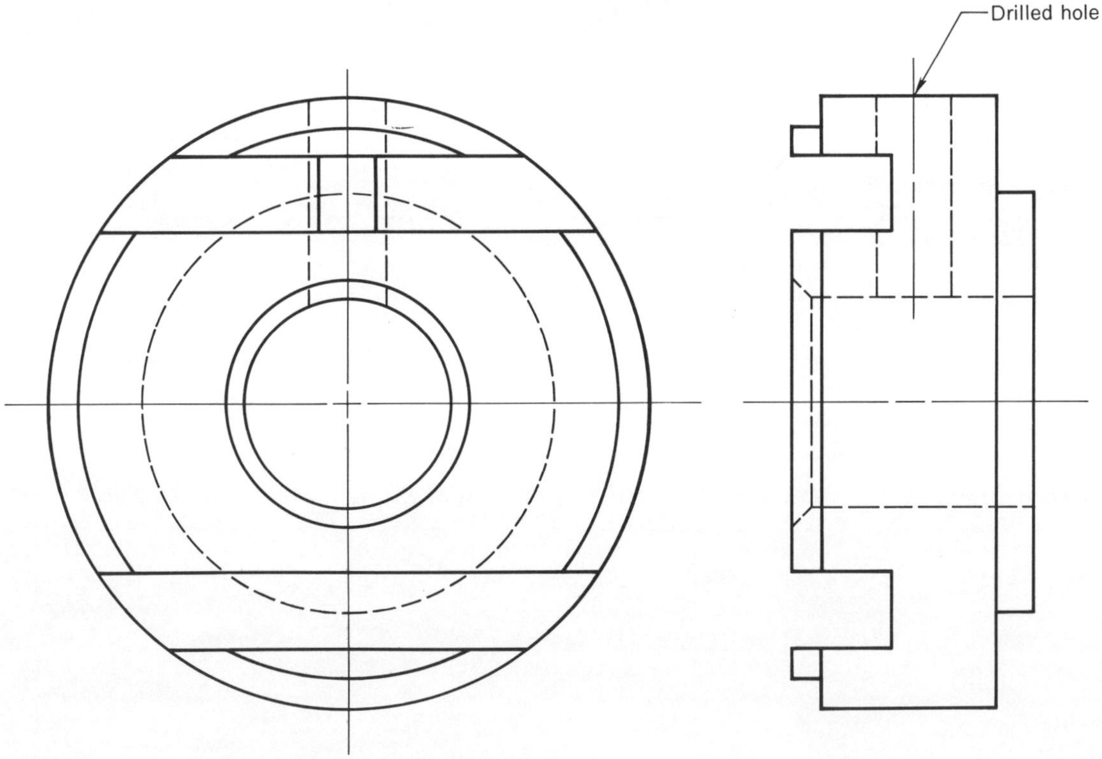
Drilled hole

MISSING VIEWS MULTIVIEW PROJECTION	DRAWN BY	FILE NO.	DRAWING 21

	DRAWN BY	FILE NO.	DRAWING

(1) SLEEVE
Sketch full section.

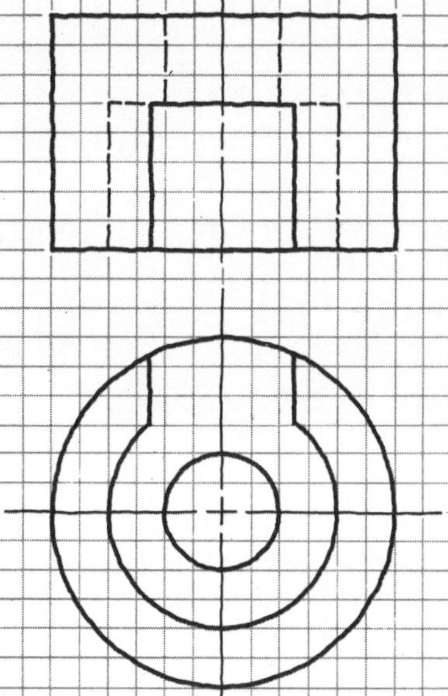

(2) LINK Sketch half section

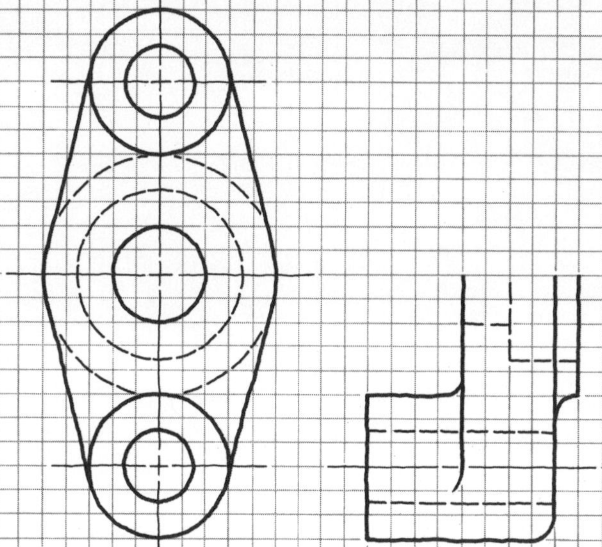

(3) PULLER BLOCK
Sketch front view.

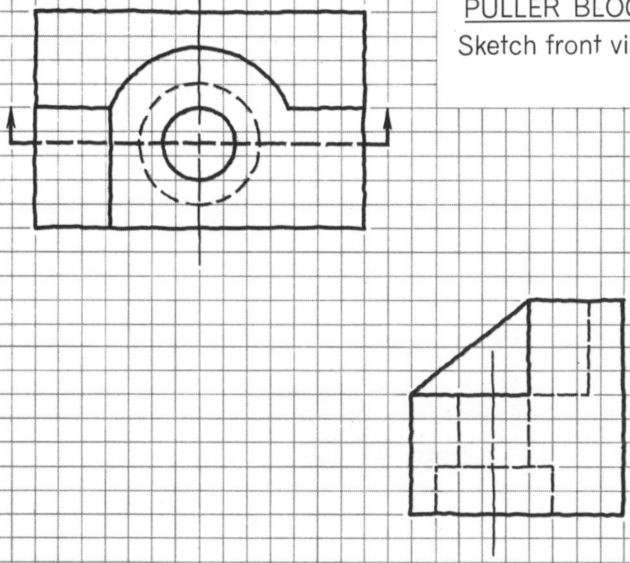

(4) CLIP STOP
Sketch full section.

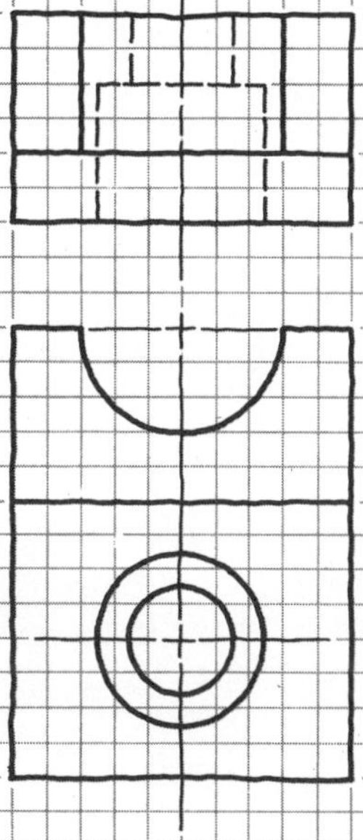

(5) CAP
Sketch half section.

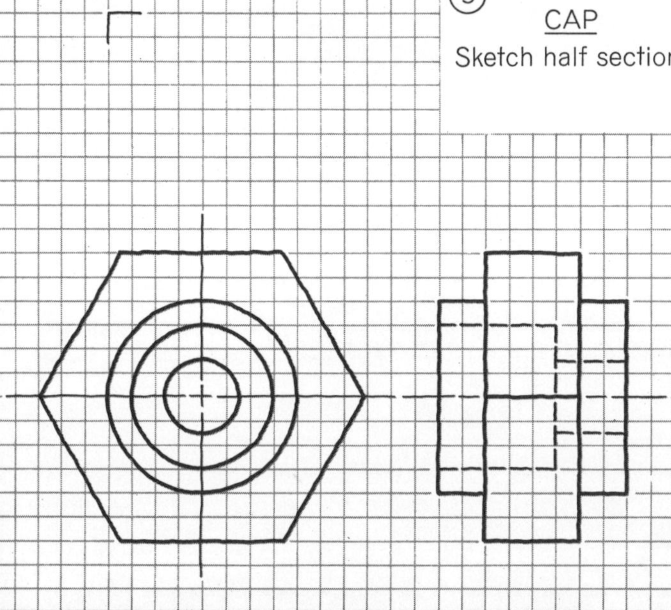

(6) COLLAR
Sketch full section.

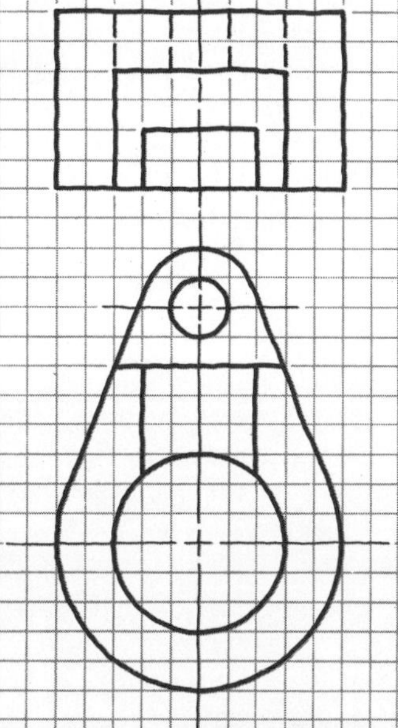

FULL AND HALF SECTIONAL VIEWS	DRAWN BY	FILE NO.	DRAWING 23

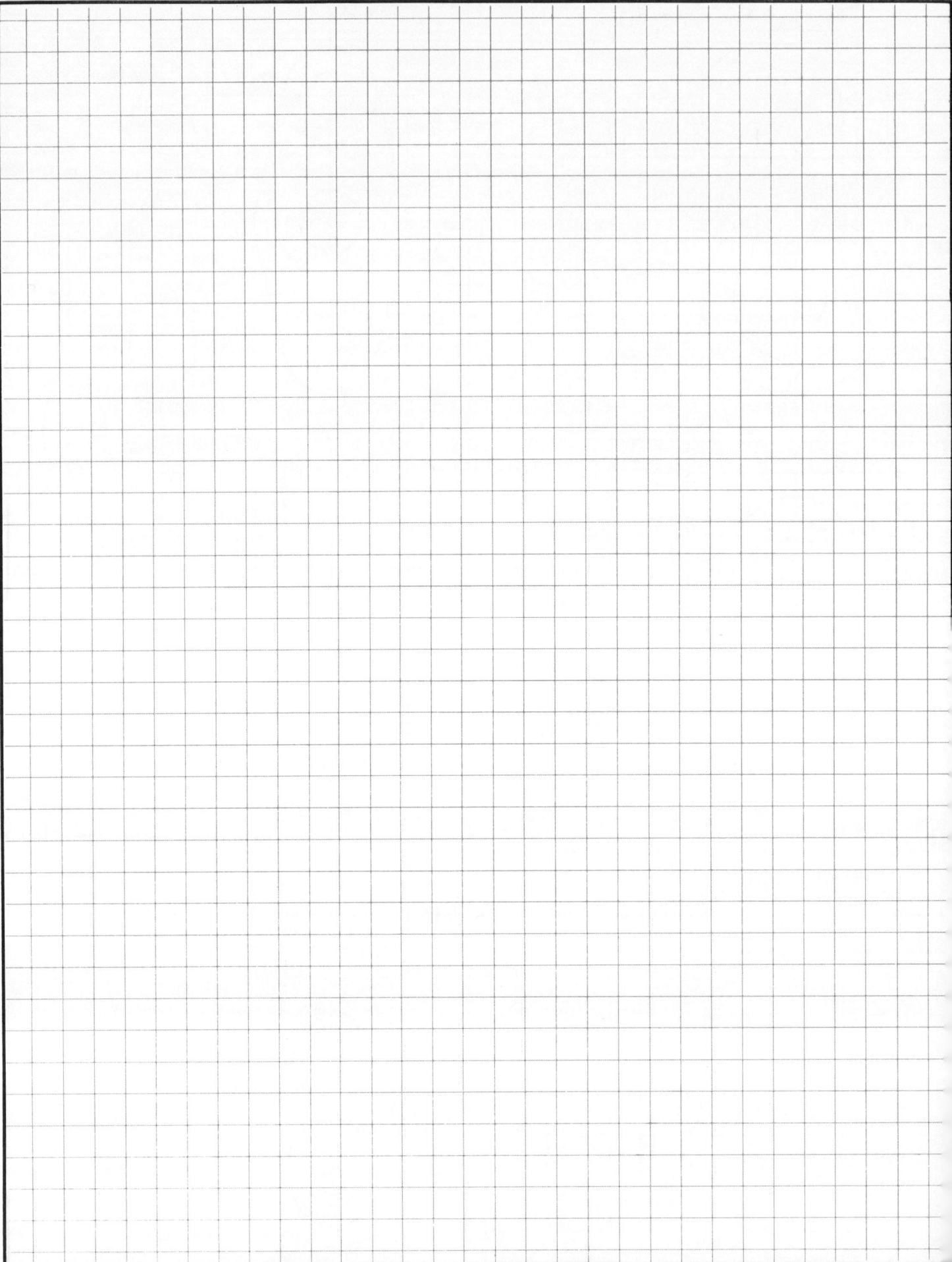

	DRAWN BY	FILE NO.	DRAWIN

(1)

SUPPORT BRACKET
Draw full section.

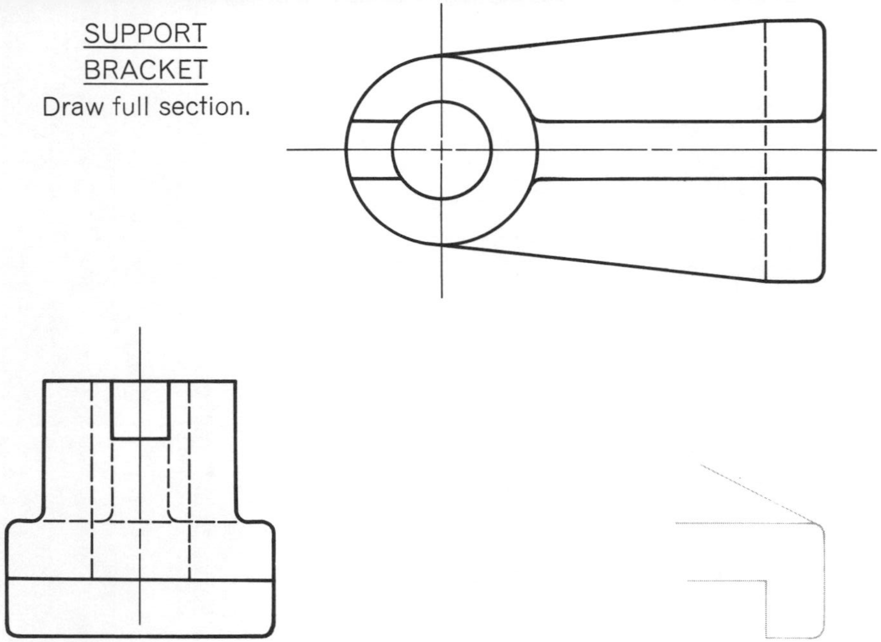

(2)

CONNECTING ARM
Draw revolved section and broken-out section.

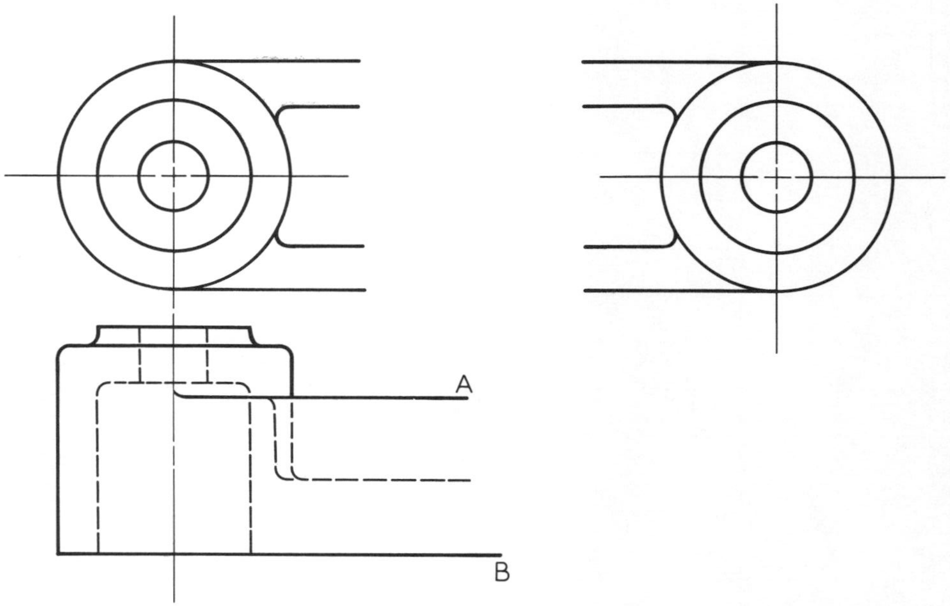
A
B

(3)

BEARING
Draw offset section to include features A, B, and C. Show cutting plane.

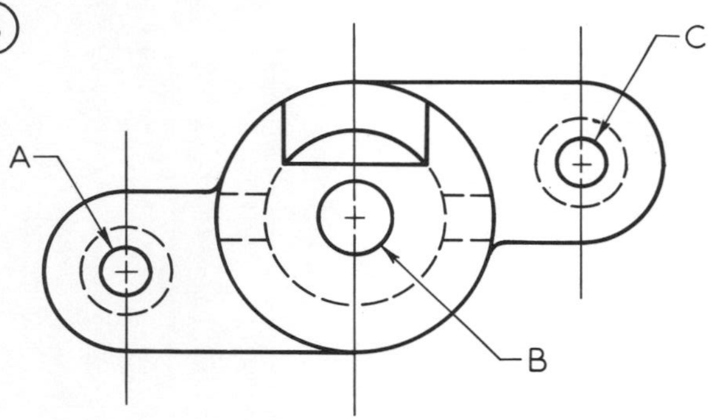
A
B
C

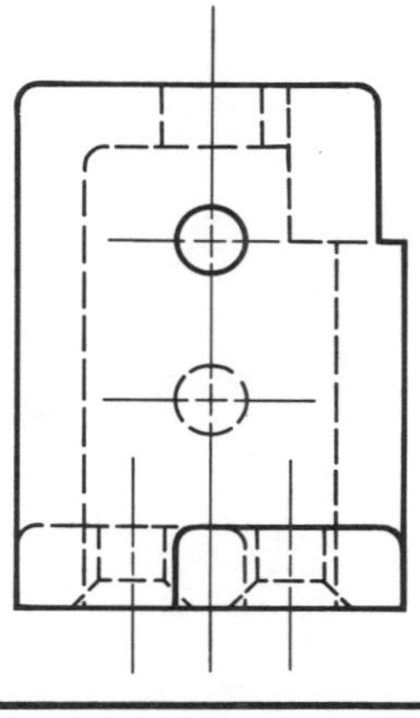

(4)

GLAND BASE
Draw full section.

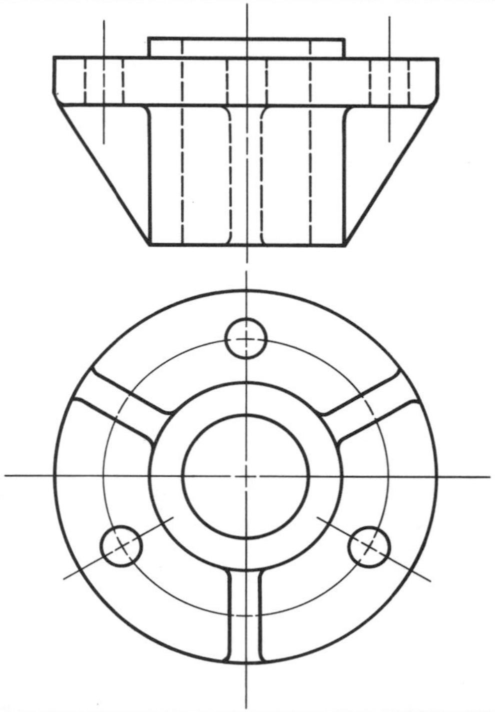

DRAWN BY

FILE NO.

DRAWN

①

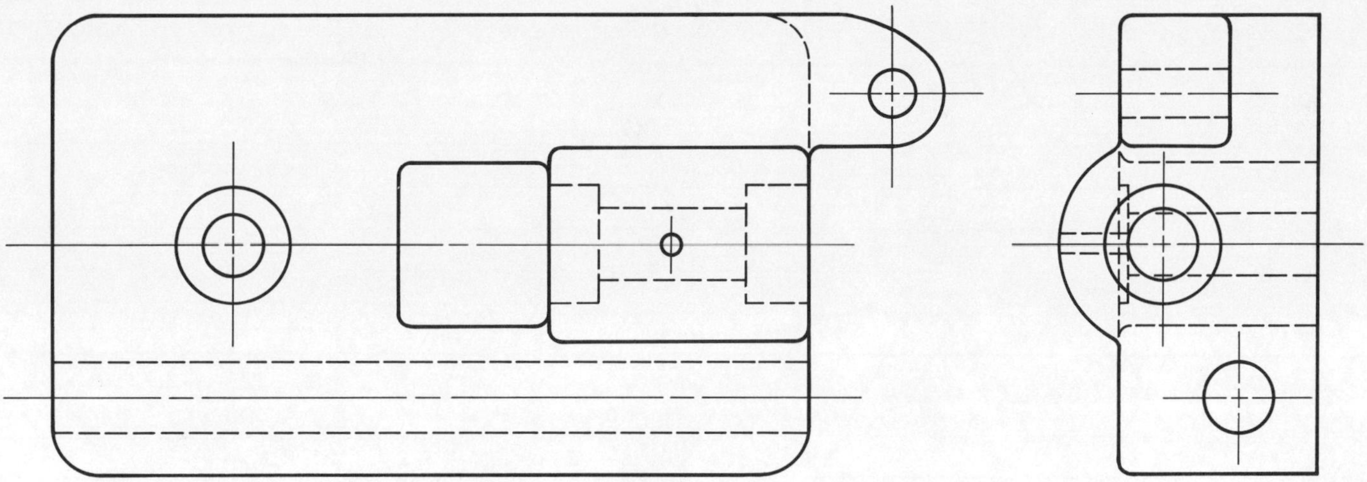

GUIDE SHOE

Draw full section.

②

SPECIAL VALVE

Draw removed sections.

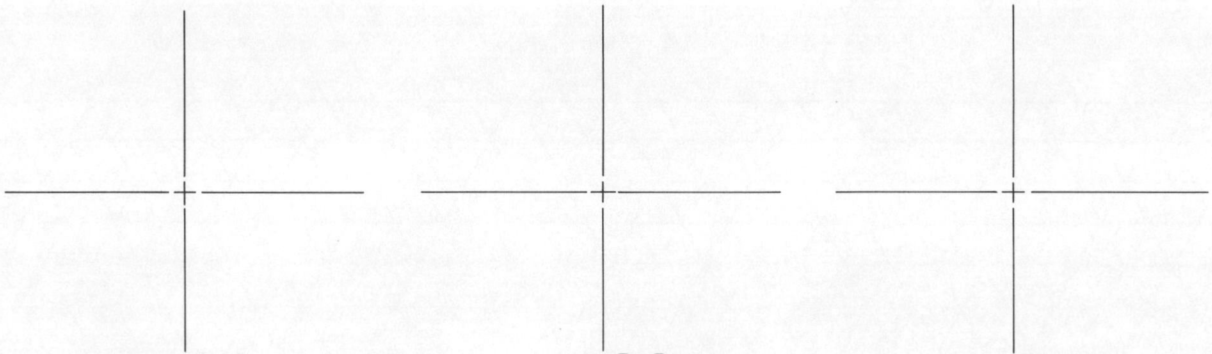

SECT A-A　　SECT B-B　　SECT C-C

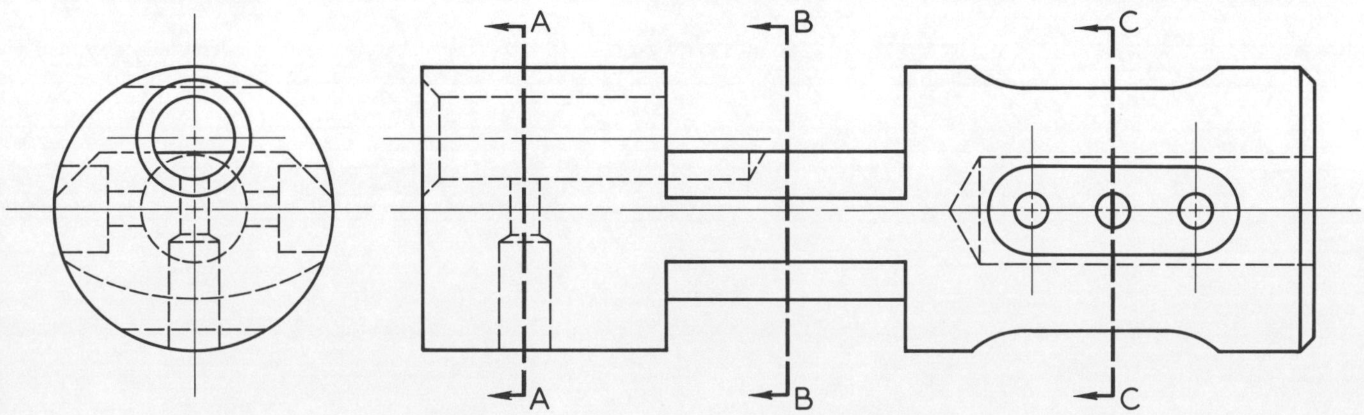
A
A
B
B
C
C

FULL AND REMOVED SECTIONAL VIEWS	DRAWN BY	FILE NO.	DRAWING 25

	DRAWN BY	FILE NO.	DRAWIN

1 CONTROL BLOCK
FOR FIXTURE

Sketch auxiliary view of surface A only.

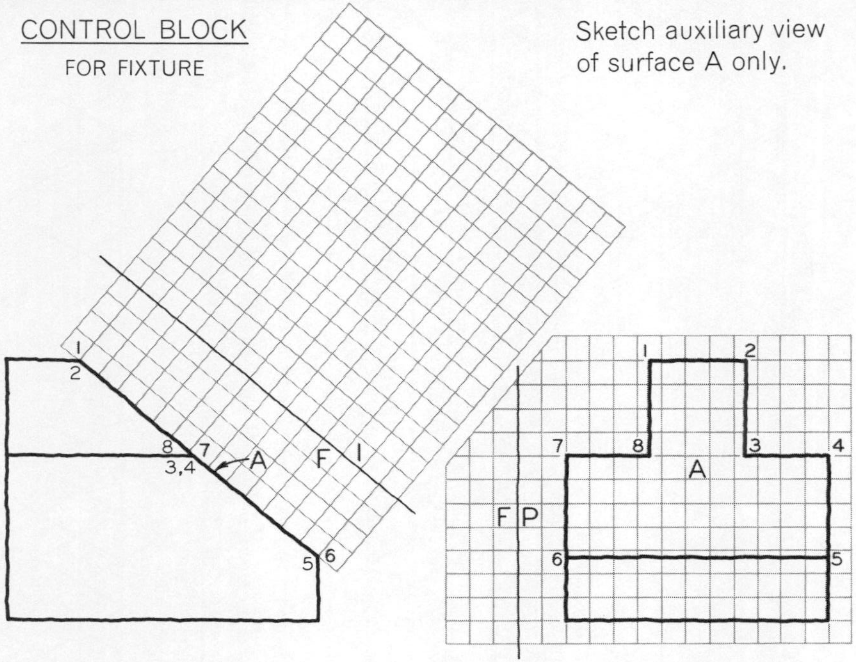

1
2
8
7
3,4
A
F
I
5
6
1
2
7
8
3
4
A
F
P
6
5

2 ANGLE BASE
FOR BOTTLE CAPPER

Sketch complete auxiliary view showing true size of surface A.

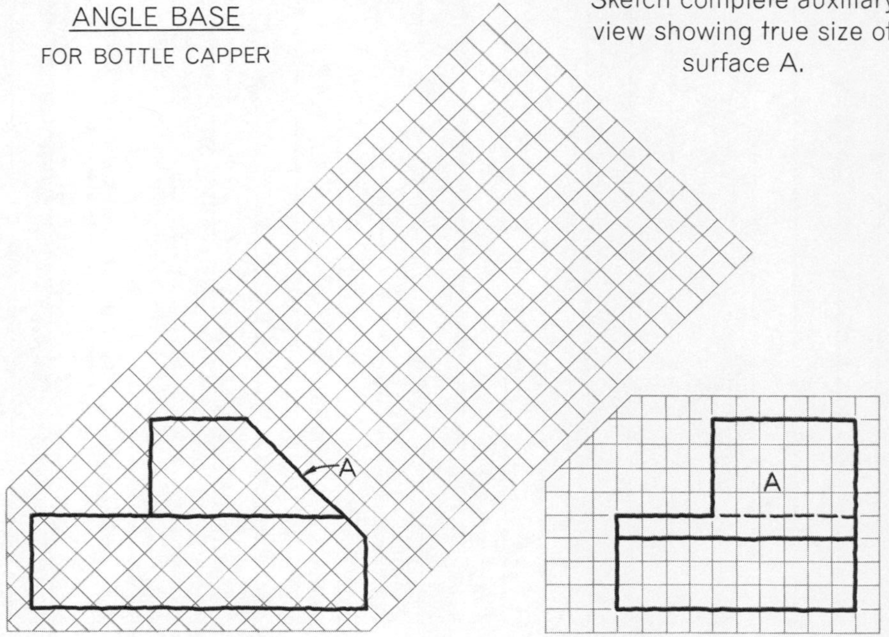

A
A

3 SHOE
FOR TRIMMER

Sketch complete auxiliary view showing true size of surface A.

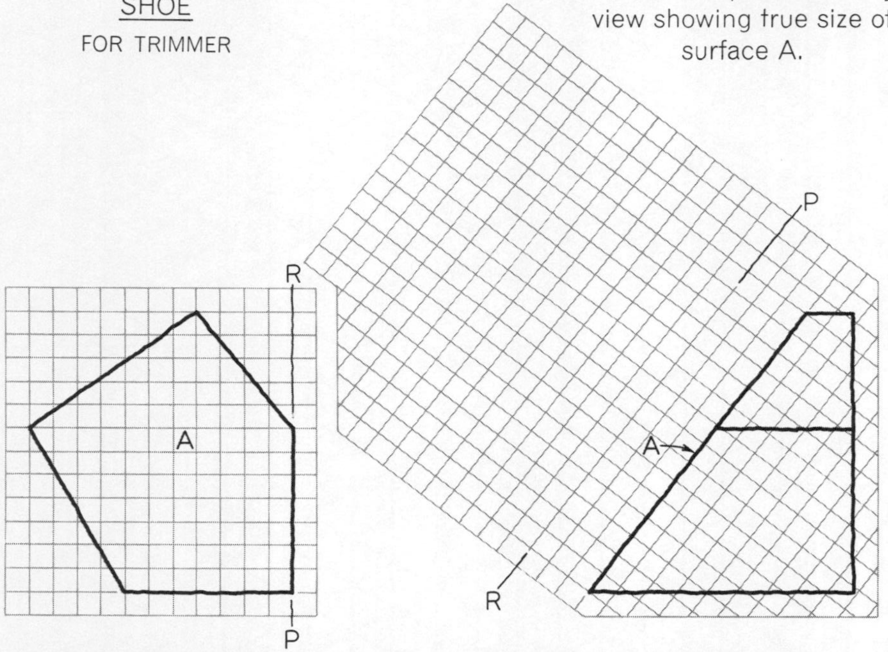

R
A
P
P
A
R

4 SLIDE PLATE
FOR DRYER

Sketch complete auxiliary view showing true angle between surfaces A and B.

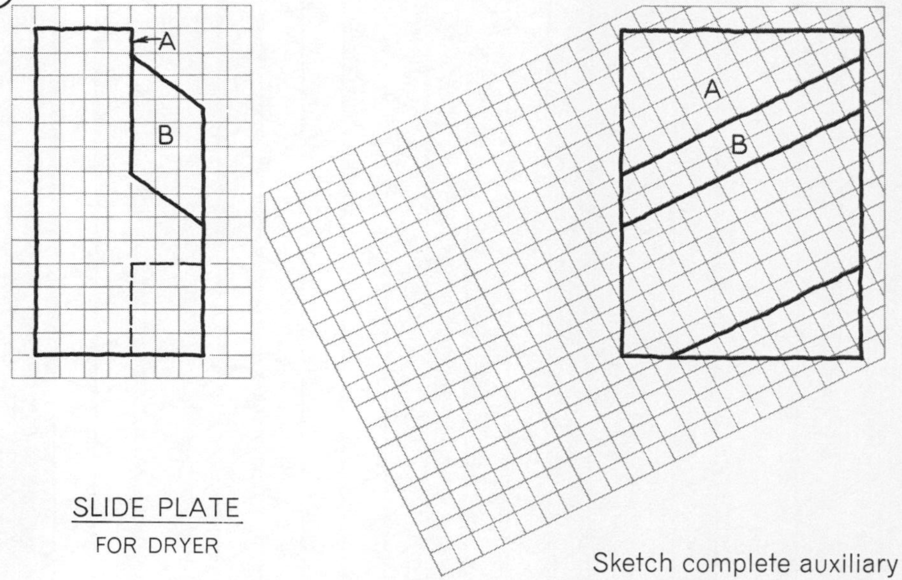

A
B
A
B

PRIMARY AUXILIARY VIEWS	DRAWN BY	FILE NO.	DRAWING 28

	DRAWN BY	FILE NO.	DRAWIN

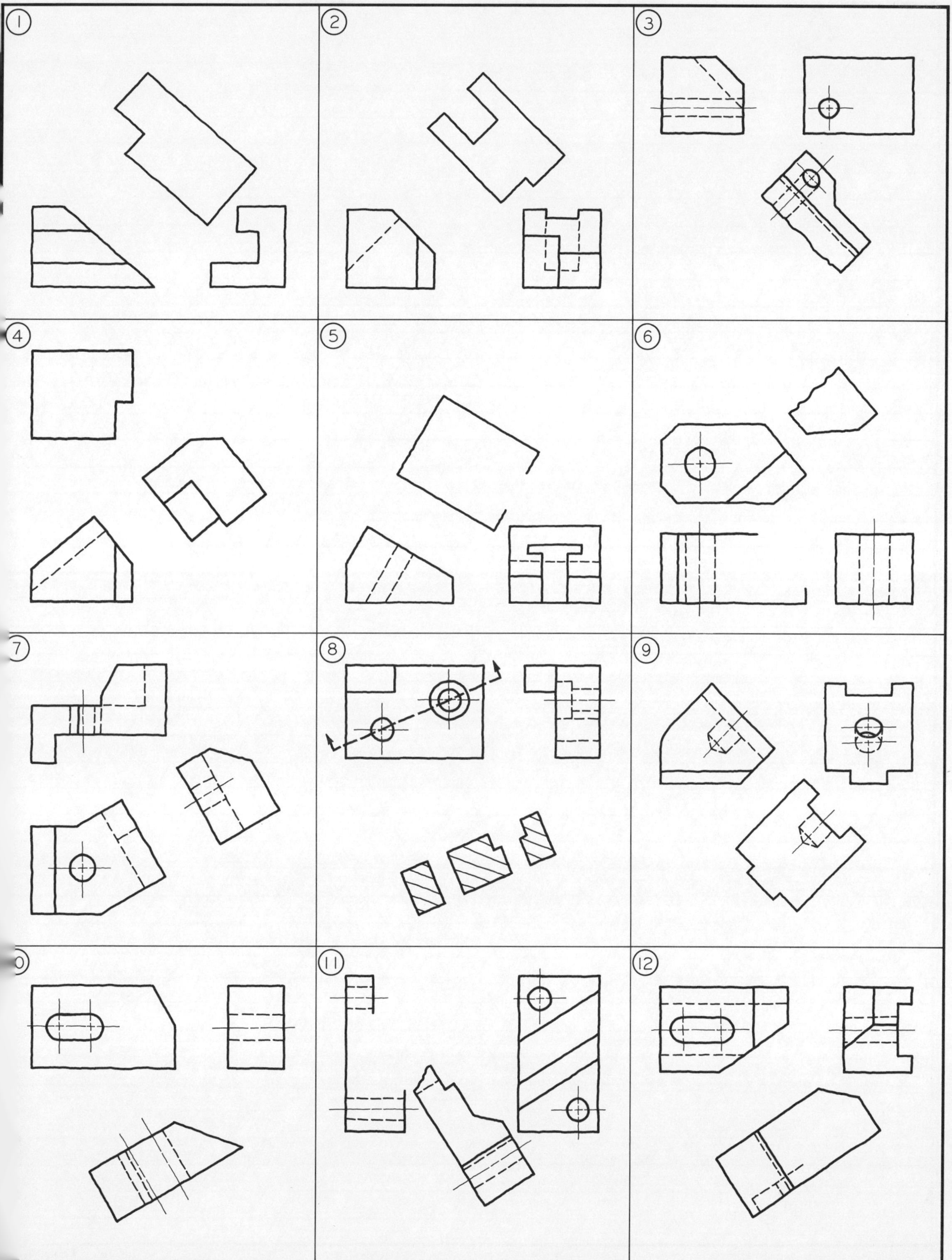

1
2
3
4
5
6
7
8
9
10
11
12

PRIMARY AUXILIARY VIEWS	DRAWN BY	FILE NO.	DRAWING 29

	DRAWN BY	FILE NO.	DRAWI

①

MOULDING

Draw complete auxiliary view showing true size of surface A.

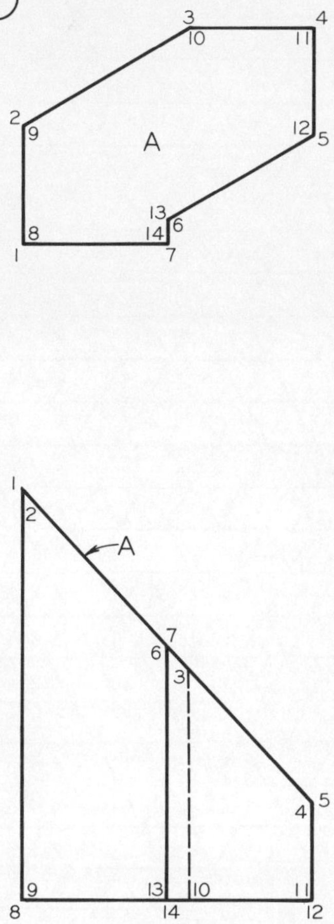

3 4 10 11
2 9
A
12 5
13 6
8 14
1 7
1
2
A
7
6
3
4 5
9 13 10 11
8 14 12

②

BEVELED SUPPORT

Draw complete auxiliary view showing true size of surface A.

A
A

③

ADJUSTABLE SLIDE

Draw complete auxiliary view showing true size of surface A.

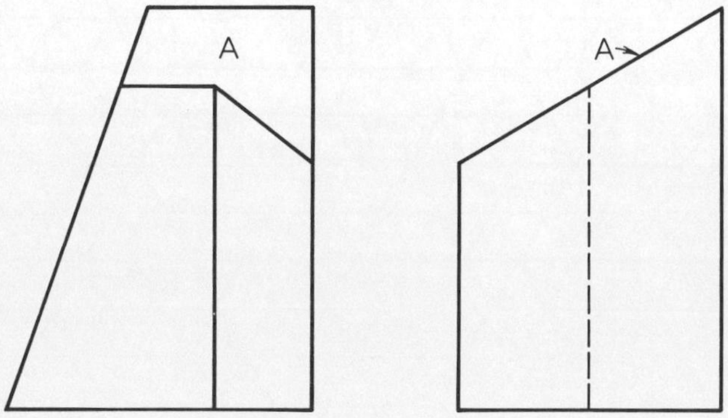

A
A

④

GUIDE BLOCK

Draw complete auxiliary view showing true angle between surfaces A and B.

A
B
B
A

PRIMARY AUXILIARY VIEWS	DRAWN BY	FILE NO.	DRAWING 30

	DRAWN BY	FILE NO.	DRAWI

1

CLIP BRACKET

Draw partial secondary auxiliary view showing true size of surface A (see detail view), and complete the given views.

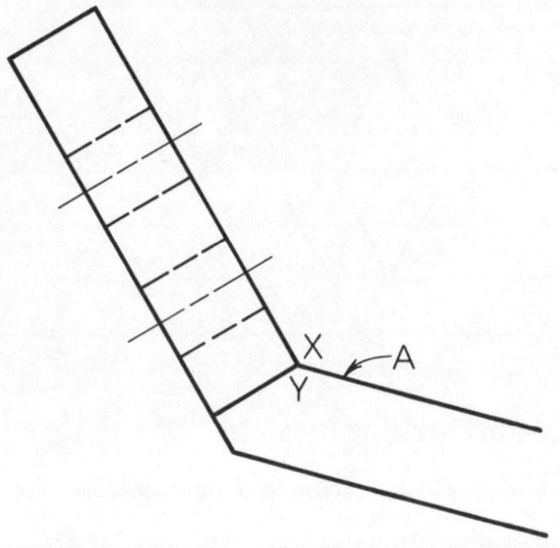

X
A
Y

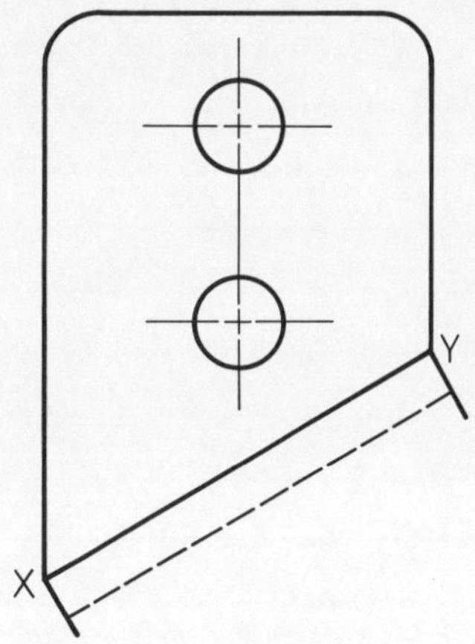

Y
X

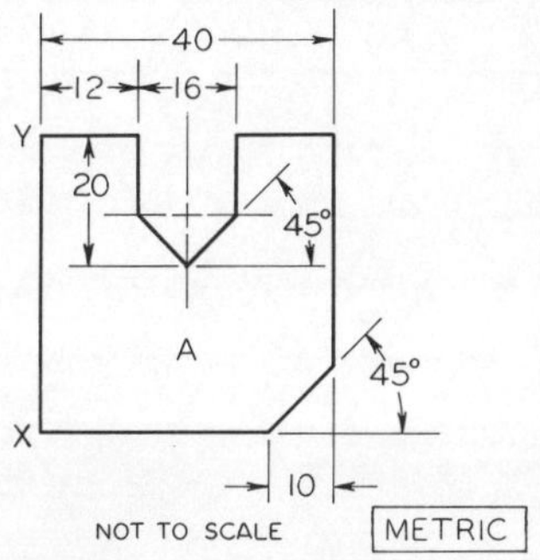

40
12
16
Y
20
45°
A
45°
X
10
NOT TO SCALE
METRIC

2

CLUTCH STOP

From given top view draw complete auxiliary views showing the true angle between surfaces A and B, and true size of surface B.

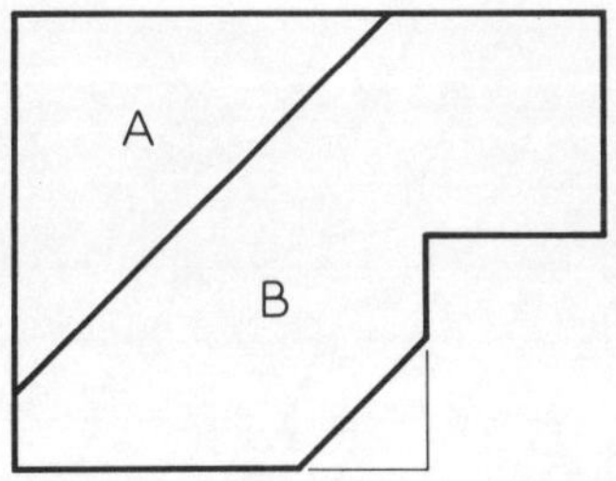

A
B

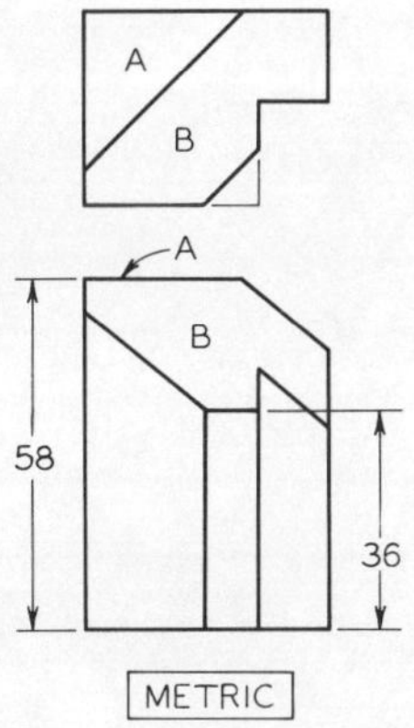

A
B
A
B
58
36
METRIC

SECONDARY AUXILIARY VIEWS	DRAWN BY	FILE NO.	DRAWING 32

	DRAWN BY	FILE NO.	DRAWIN

(1) Revolve point 3 to the extreme right of axis 1-2.
Measure the angle of revolution.

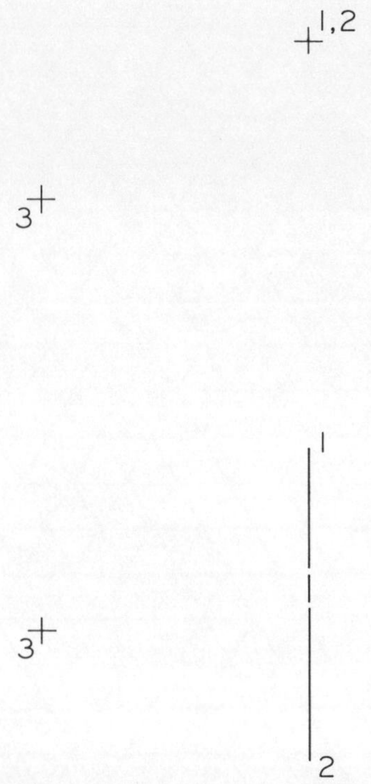
1,2
3
1
3
2

(2) Revolve line 3-4 to a horizontal position above axis 1-2.

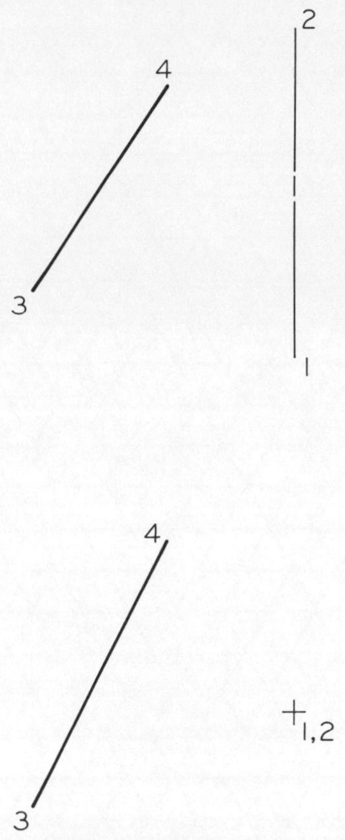
2
4
3
1
4
1,2
3

(3) Revolve the prism about edge 2-3 until surface 1-2-3-4-5 is parallel to the frontal plane of projection.

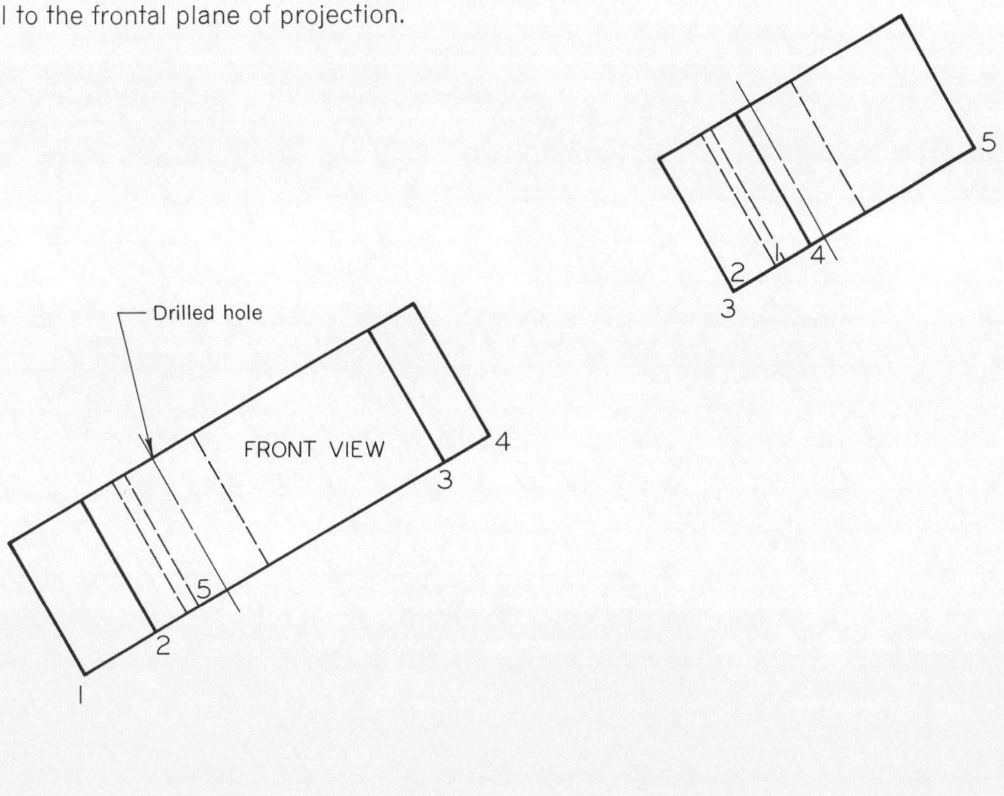
5
4
2
3
Drilled hole
FRONT VIEW
4
3
5
2
1

	DRAWN BY	FILE NO.	DRAWIN

1. STOP BLOCK

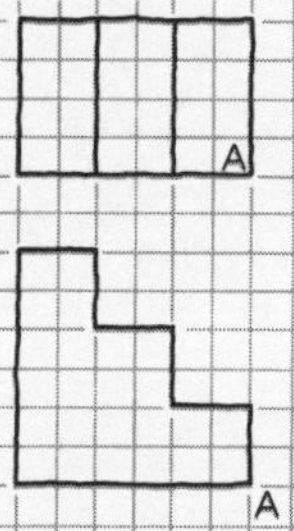
A
A

2. JIG GUIDE

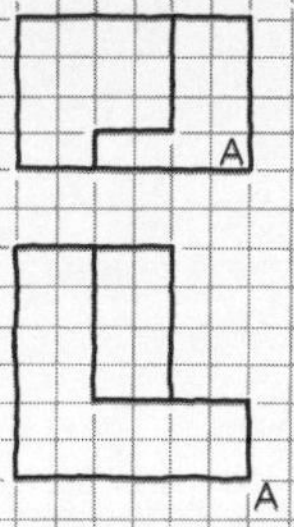
A
A

3. CHUCK JAW

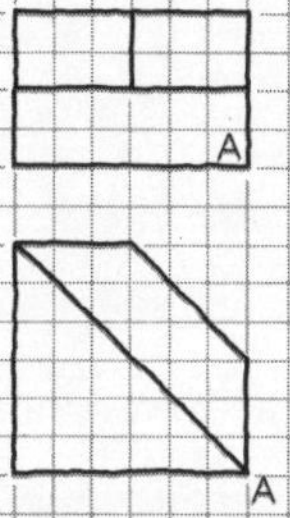
A
A

4. CAM STOP

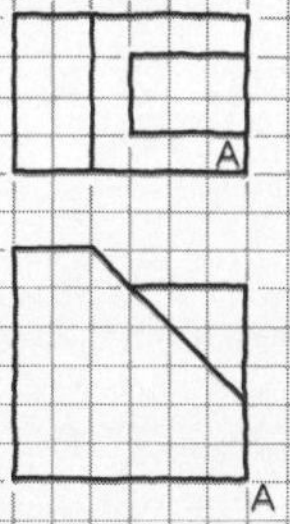
A
A

. GUIDE

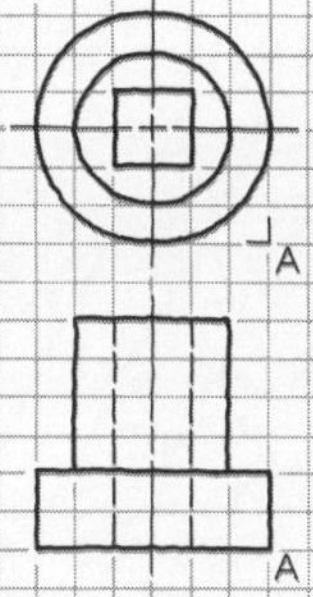
A
A

. BAR SUPPORT

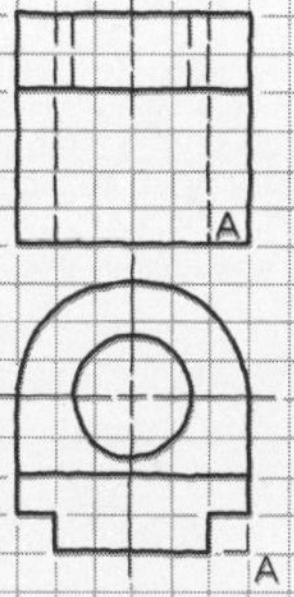
A
A

(1) A

(2) A

(3) A

(4) A

(5) A

(6) A

FREEHAND ISOMETRIC DRAWING	DRAWN BY	FILE NO.	DRAWING 34

	DRAWN BY	FILE NO.	DRAWIN

1

GUIDE BASE

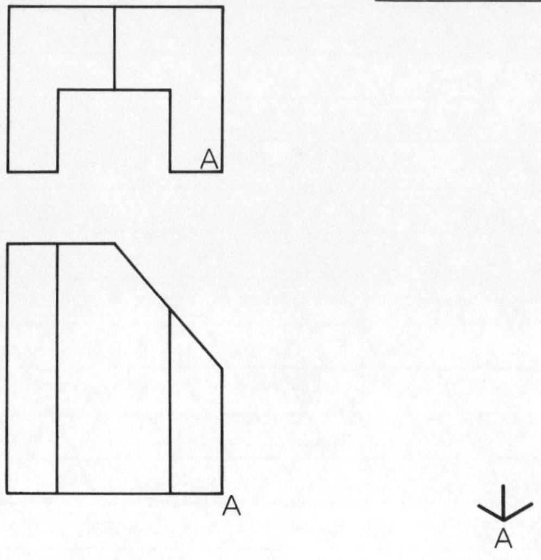

A
A
A

2

TRIP BLOCK

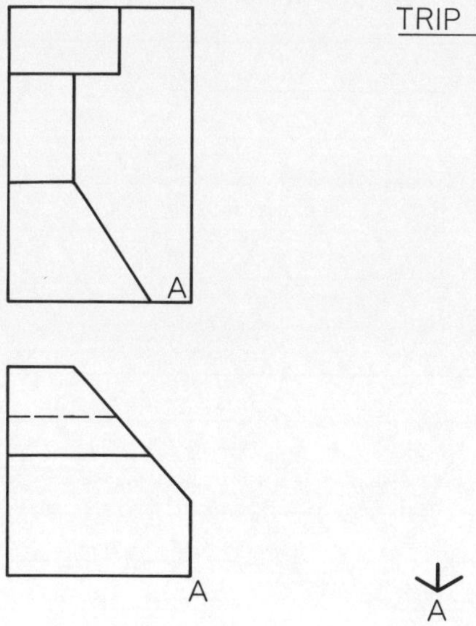

A
A
A

3

HEXAGON PLUG

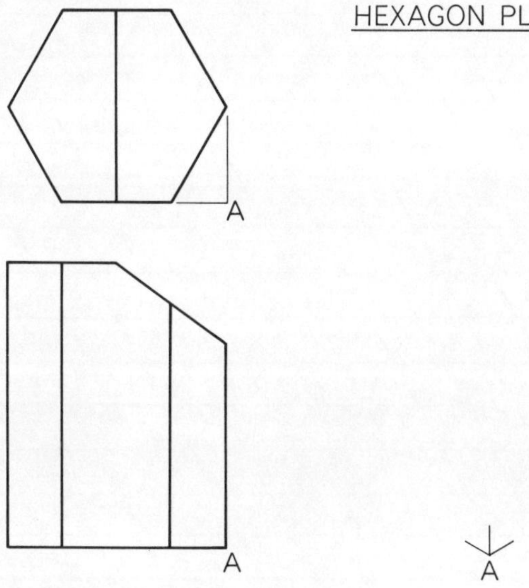

A
A
A

4

CENTERING JAW

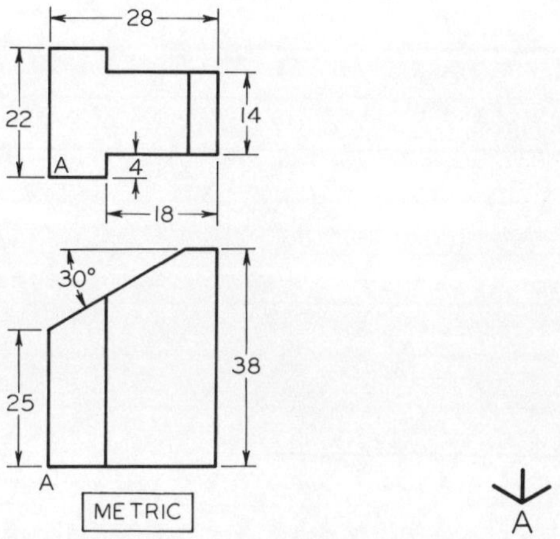

28
22
14
A
4
18
30°
38
25
A
METRIC
A

5

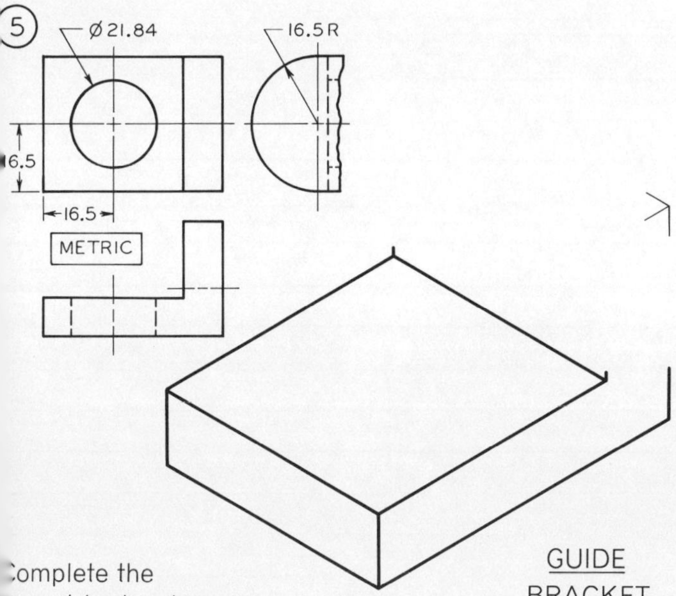

Ø21.84
16.5 R
16.5
16.5
METRIC

Complete the isometric drawing.

GUIDE BRACKET

6

BRACE

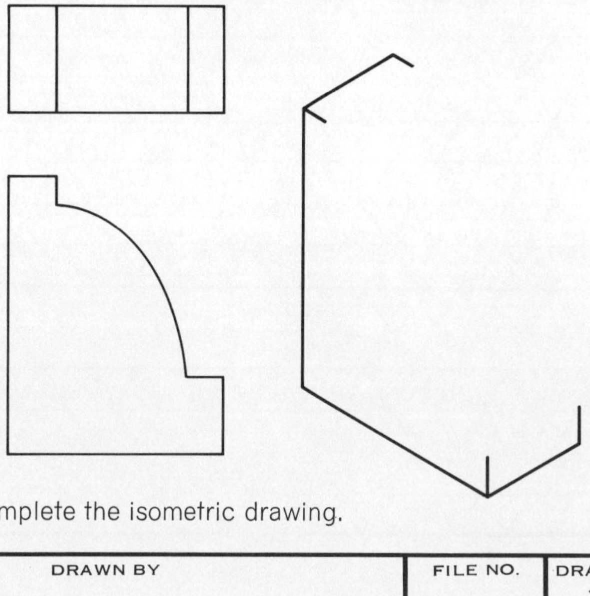

Complete the isometric drawing.

MECHANICAL ISOMETRIC DRAWING	DRAWN BY	FILE NO.	DRAWING 35

	DRAWN BY	FILE NO.	DRAWI

(1)

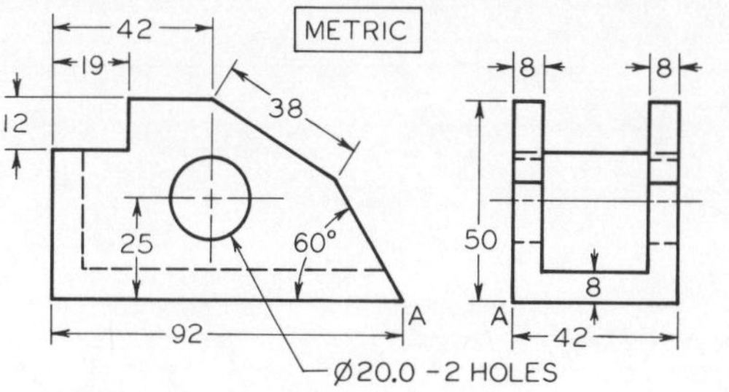
METRIC
42
19
12
38
25
60°
92
A
Ø20.0 -2 HOLES
8
8
50
8
A
42

<u>GUIDE SHOE</u>

Draw isometric drawing.

A

(2)

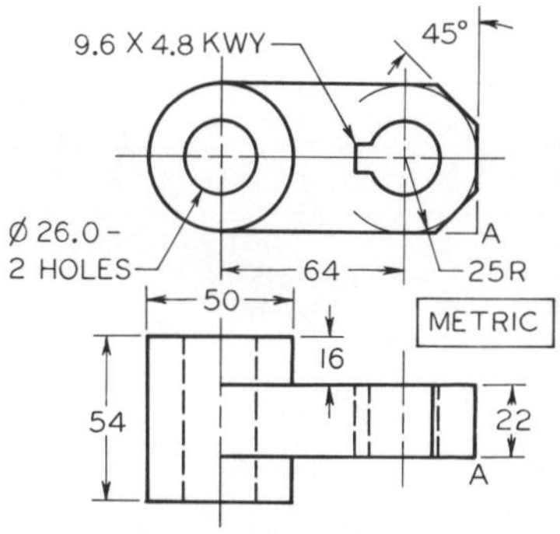
9.6 X 4.8 KWY
45°
Ø 26.0 -
2 HOLES
A
64
25R
50
METRIC
16
54
22
A

<u>TRIP LEVER</u>

Draw isometric drawing.

A

MECHANICAL ISOMETRIC DRAWING	DRAWN BY	FILE NO.	DRAWING 36

	DRAWN BY	FILE NO.	DRAWIN

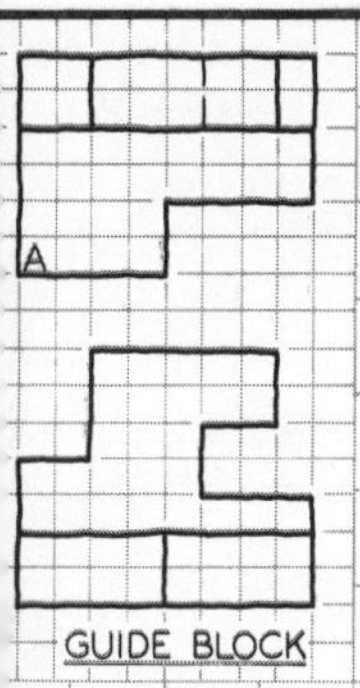
A
GUIDE BLOCK

1

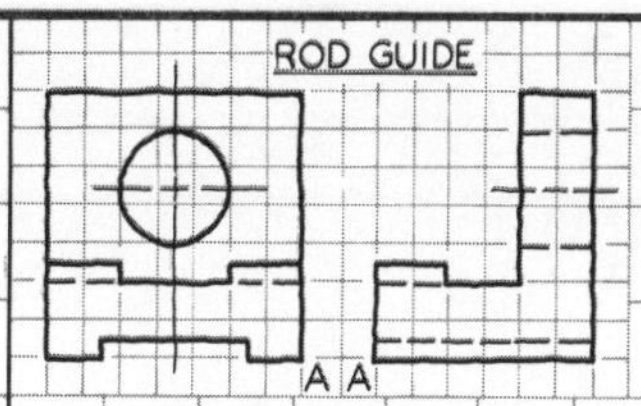
ROD GUIDE
A A

2

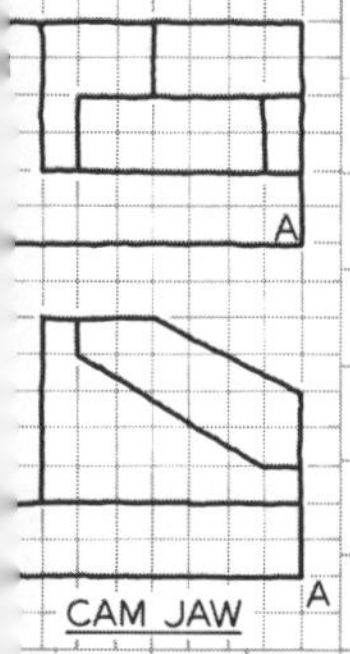
A
CAM JAW
A

3

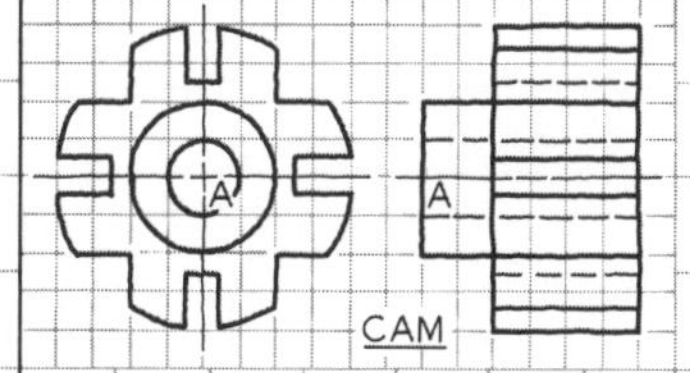
A
A
CAM

4

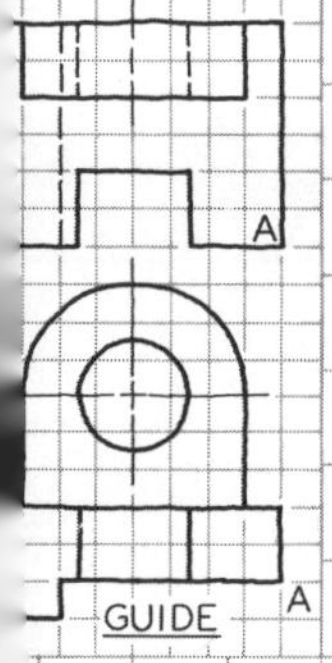
A
GUIDE
A

5

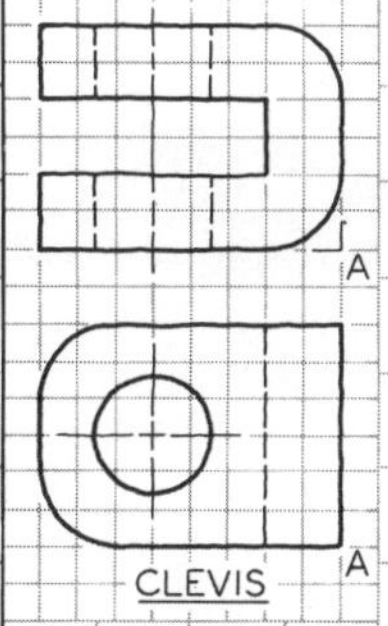
A
CLEVIS
A

6

FREEHAND OBLIQUE PROJECTION	DRAWN BY	FILE NO.	DRAWING 38

	DRAWN BY	FILE NO.	DRAW

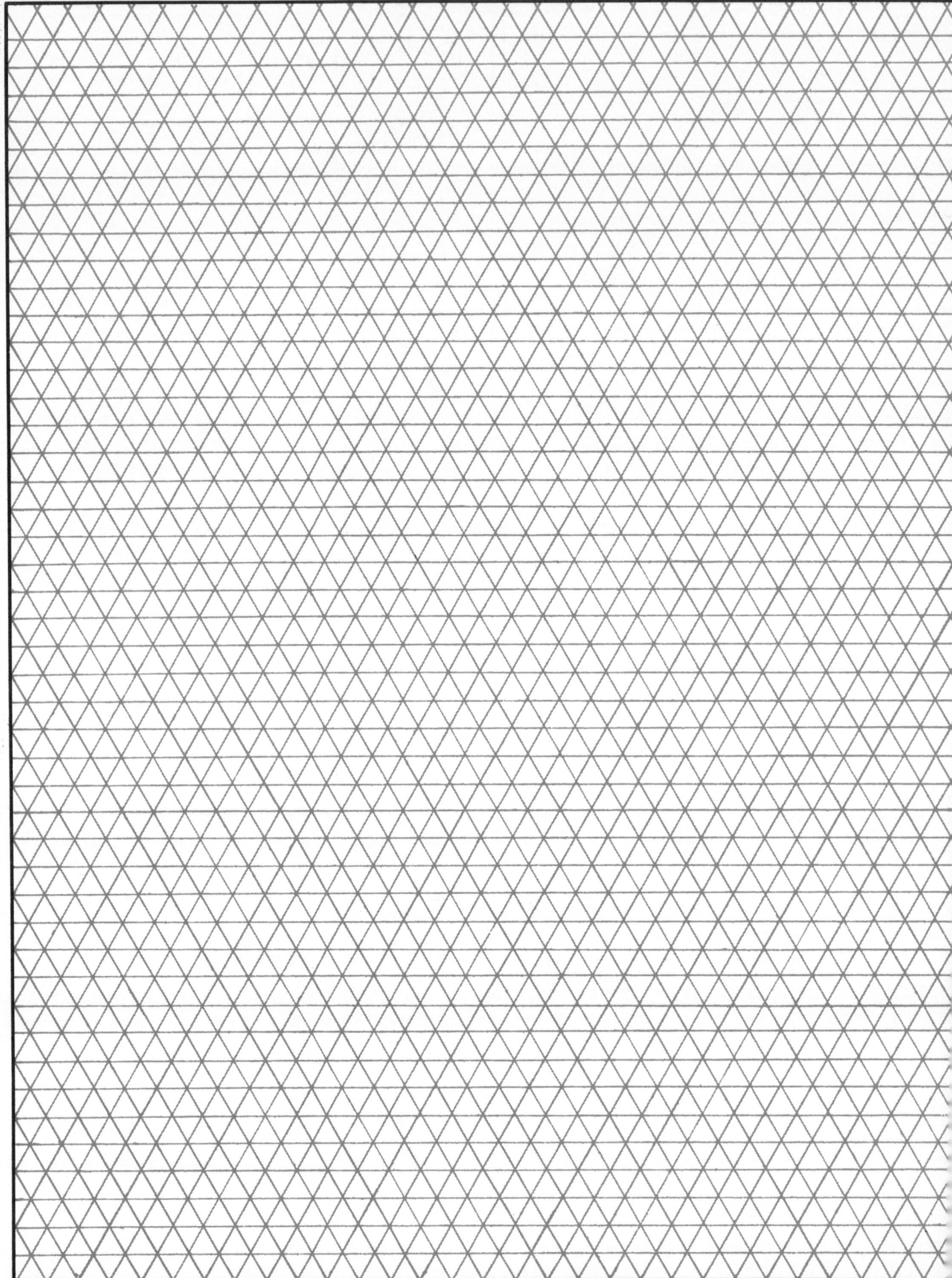

(1)

LOCK PLATE

Draw cavalier drawing.

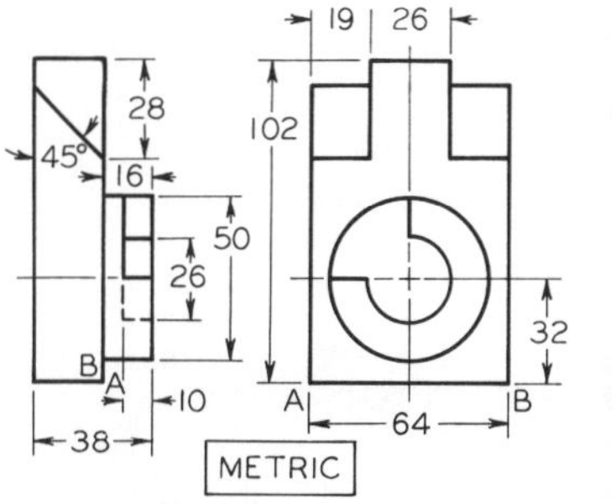

19
26
28
45°
16
102
50
26
32
B
A
10
A
64
B
38
METRIC

(2)

TOGGLE LINK

Draw cabinet drawing.

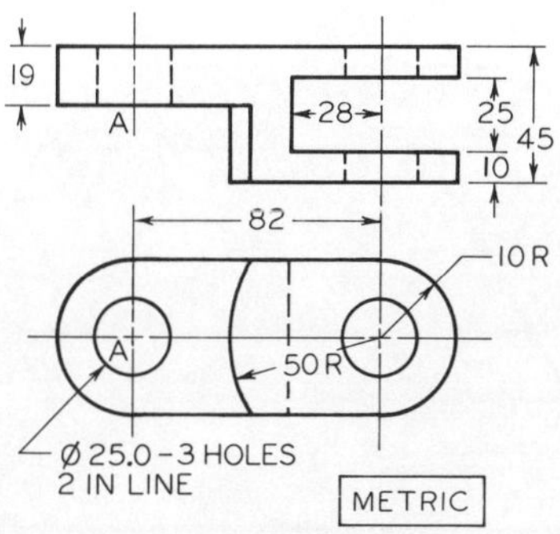

19
A
28
25
45
10
82
10 R
A
50 R
Ø 25.0 - 3 HOLES
2 IN LINE
METRIC

	DRAWN BY	FILE NO.	DRAW

(1) Add dimensions freehand.

BLANK

FOR SPECIAL NUT
CRS — 1 REQD
SCALE: FULL SIZE

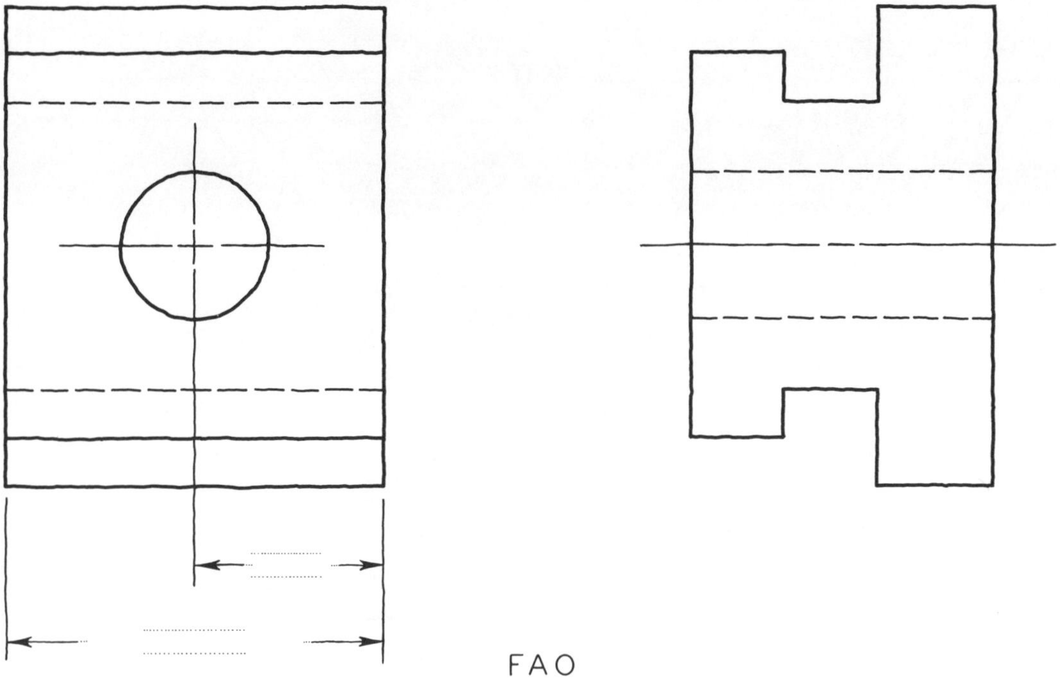

FAO

(2) LINK

FOR PRINTING PRESS
CI — 2 REQD
SCALE: HALF SIZE

Add dimensions freehand.

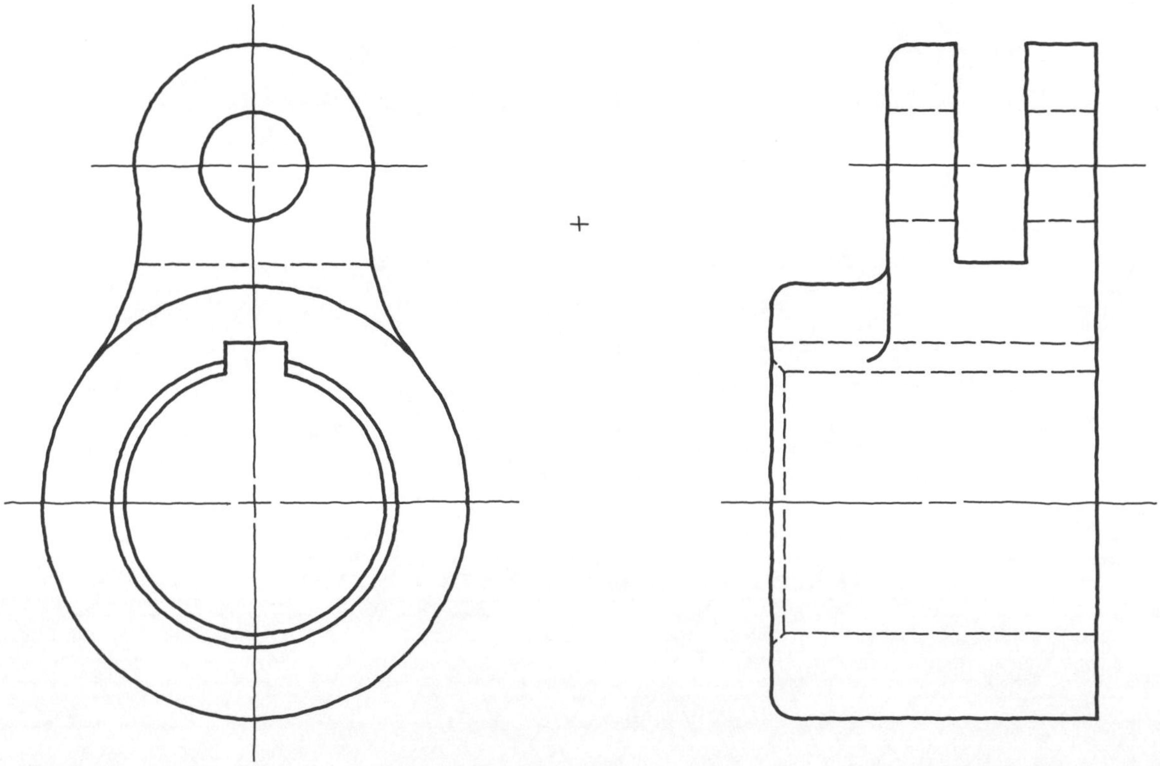

FREEHAND DIMENSIONING	DRAWN BY	FILE NO.	DRAWING 40

	DRAWN BY	FILE NO.	DRAWING

①

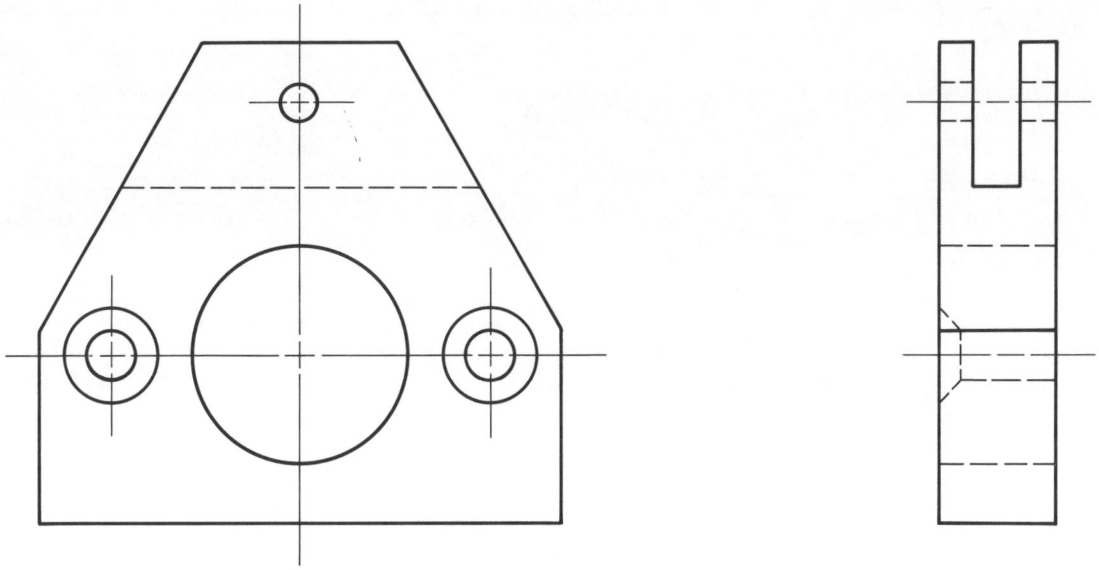

CARRIER

FOR RADIAL BEARING
CRS — 4 REQD
FULL SIZE

Add dimensions mechanically.

②

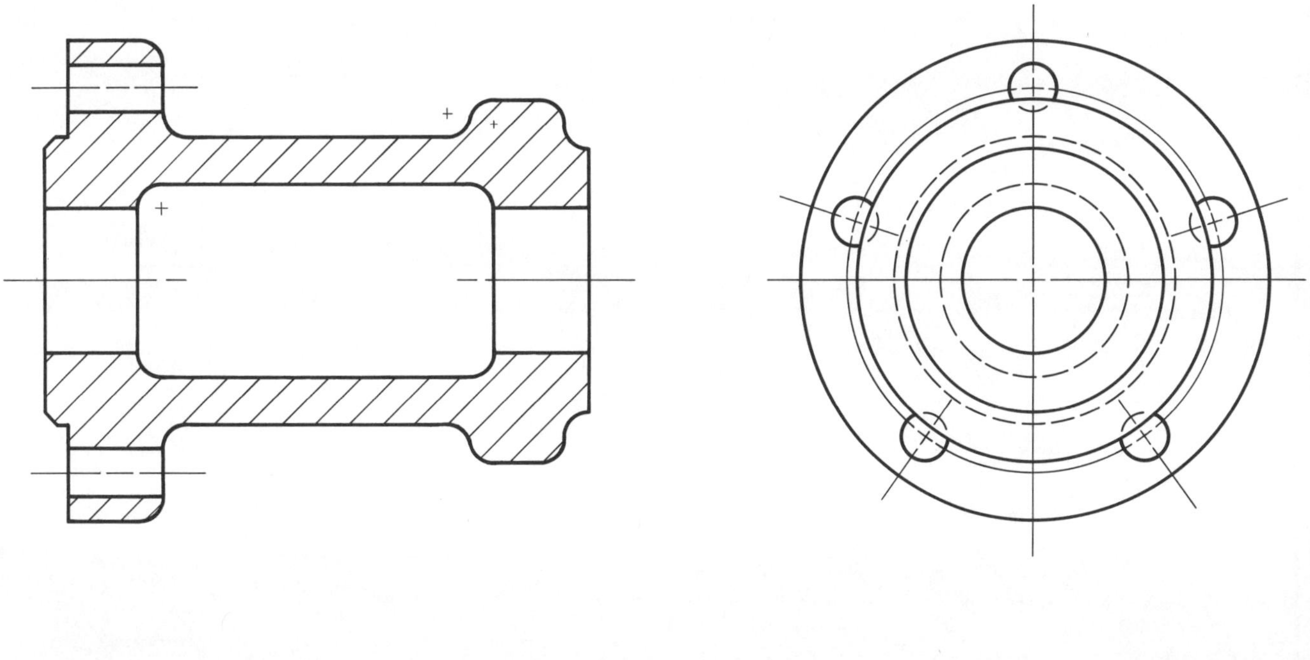

SUPPORT FRAME

FOR PRESS
CI — I REQD
FULL SIZE

Add dimensions mechanically.

MECHANICAL DIMENSIONING	DRAWN BY	FILE NO.	DRAWING 41

	DRAWN BY	FILE NO.	DRAW

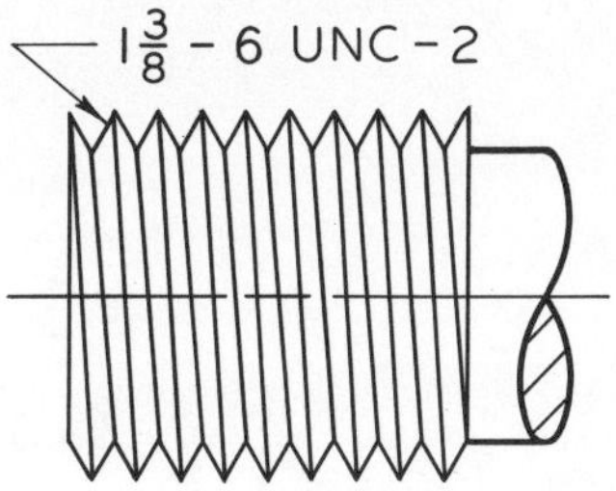
$1\frac{3}{8}$ - 6 UNC - 2

FIG. I

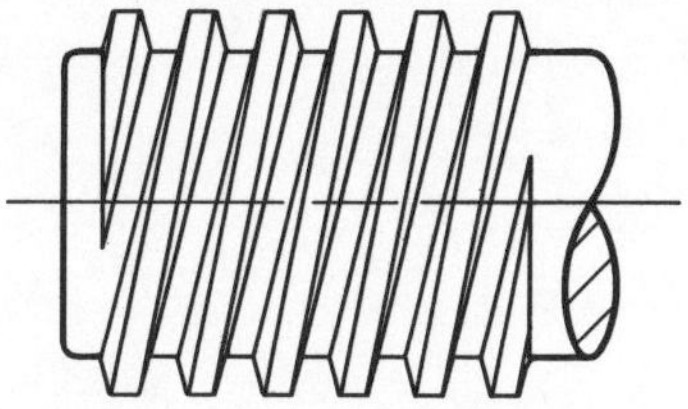

FULL SCALE

FIG. II

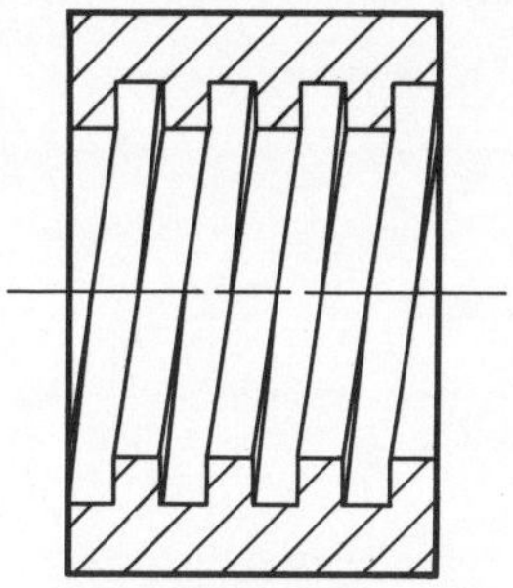

SCALE: 1 = 2

FIG. III

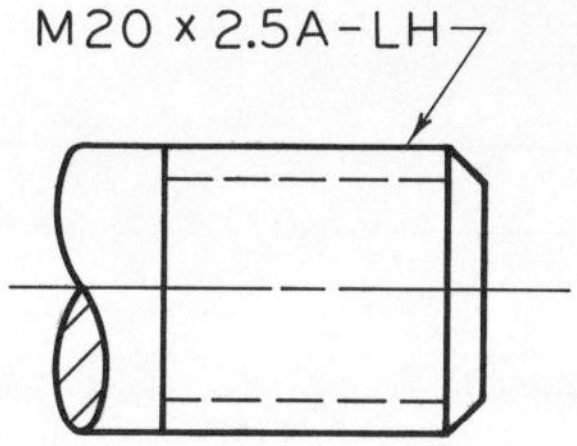
M20 x 2.5A-LH

FIG. IV

LETTER THE CORRECT INFORMATION FOR THE ABOVE FIGURES IN THE SPACES PROVIDED.

	FIG. I	FIG. II	FIG. III	FIG. IV
1. Number of threads per inch				
2. RH or LH thread				
3. Thread form used				
4. Included angle of thread				
5. Single or which multiple thread				
6. Pitch of thread				
7. Lead of thread				
8. Major diameter				
9. Internal or external thread				
10. Section or elevation view				
11. Method of representation				

IDENTIFY THE FOLLOWING

OMIT SIZE SPECIFICATIONS

A

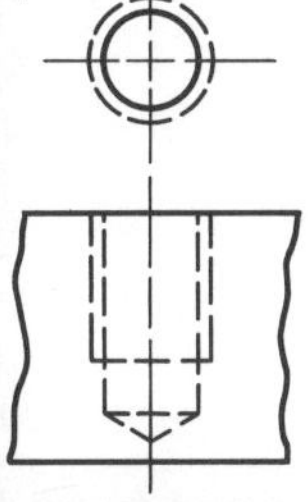

B

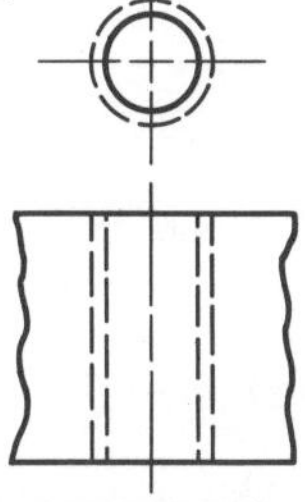

C

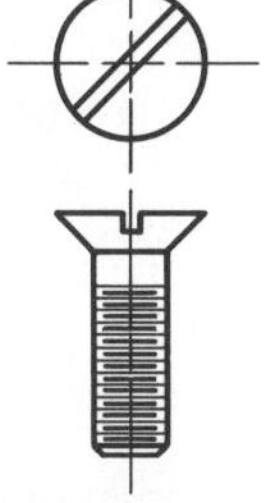

D

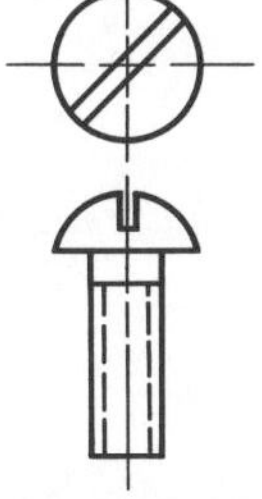

E

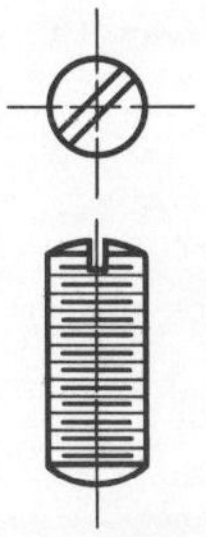

F

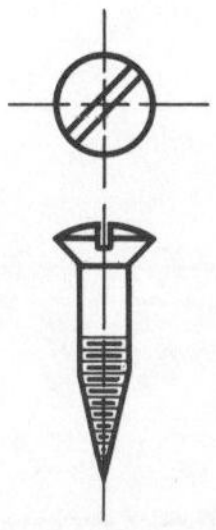

G

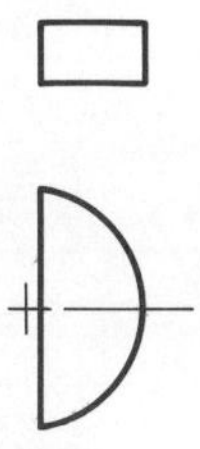

H

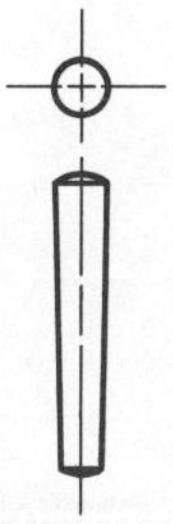

A	
B	
C	
D	
E	
F	
G	
H	

NOMENCLATURE & IDENTIFICATION THREADS AND FASTENERS	DRAWN BY	FILE NO.	DRAWING 44

	DRAWN BY	FILE NO.	DRAWIN

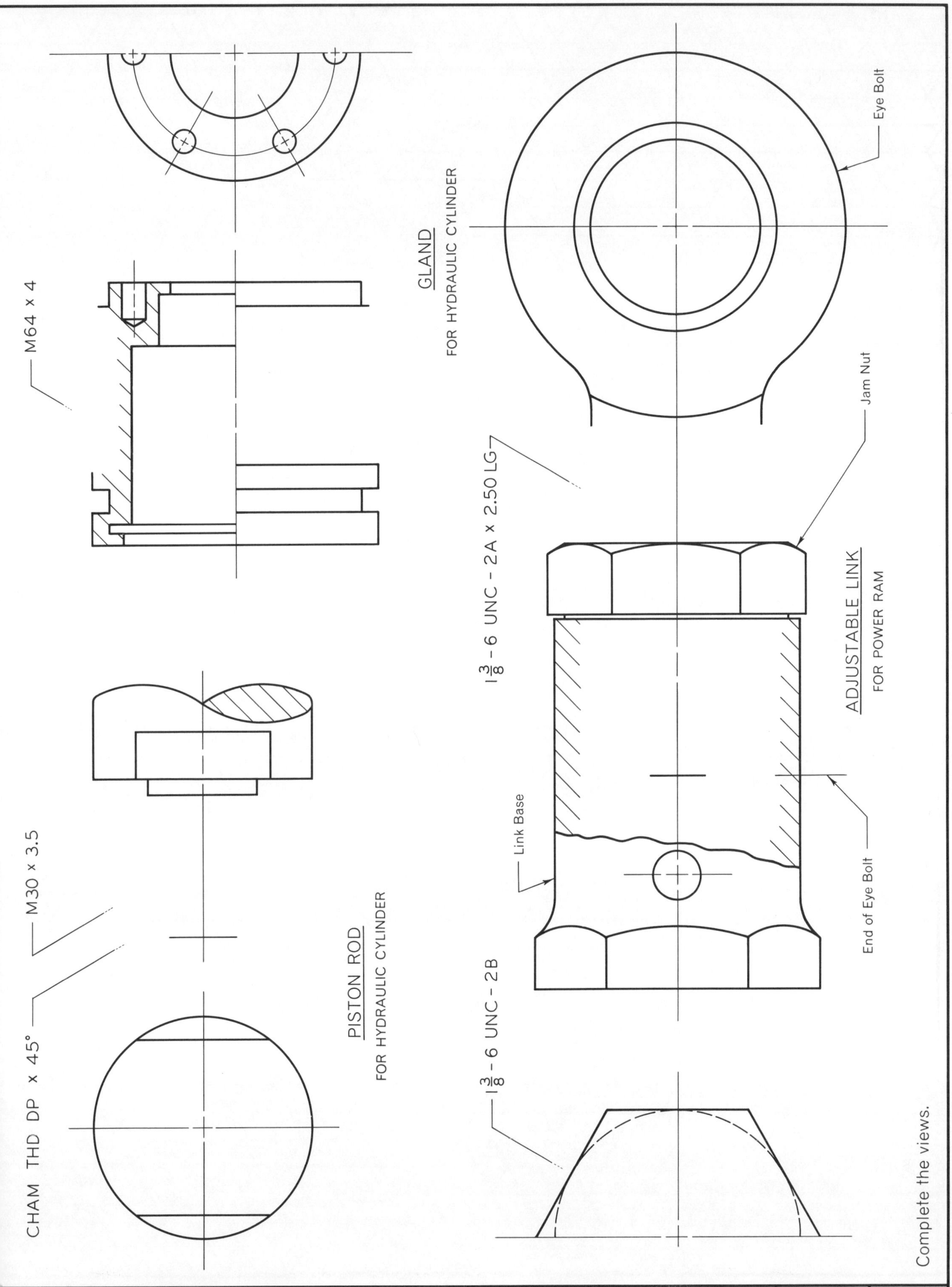
CHAM THD DP x 45°
M30 x 3.5
PISTON ROD
FOR HYDRAULIC CYLINDER
M64 x 4
GLAND
FOR HYDRAULIC CYLINDER
$1\frac{3}{8}$ - 6 UNC - 2B
$1\frac{3}{8}$ - 6 UNC - 2A x 2.50 LG
Link Base
ADJUSTABLE LINK
FOR POWER RAM
End of Eye Bolt
Jam Nut
Eye Bolt
Complete the views.

DETAILED UNIFIED THREADS	DRAWN BY	FILE NO.	DRAWING 45

	DRAWN BY	FILE NO.	DRAWIN

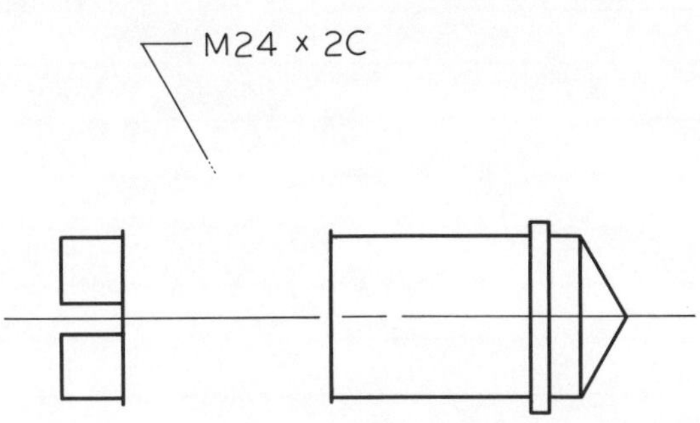
M24 × 2C

1. METERING SCREW
Complete the view.

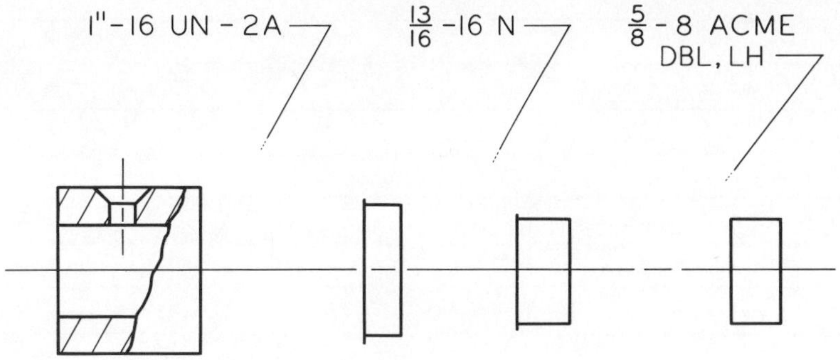
1"–16 UN – 2A
$\frac{13}{16}$–16 N
$\frac{5}{8}$ – 8 ACME
DBL, LH

2. VALVE SPINDLE
Complete the view.

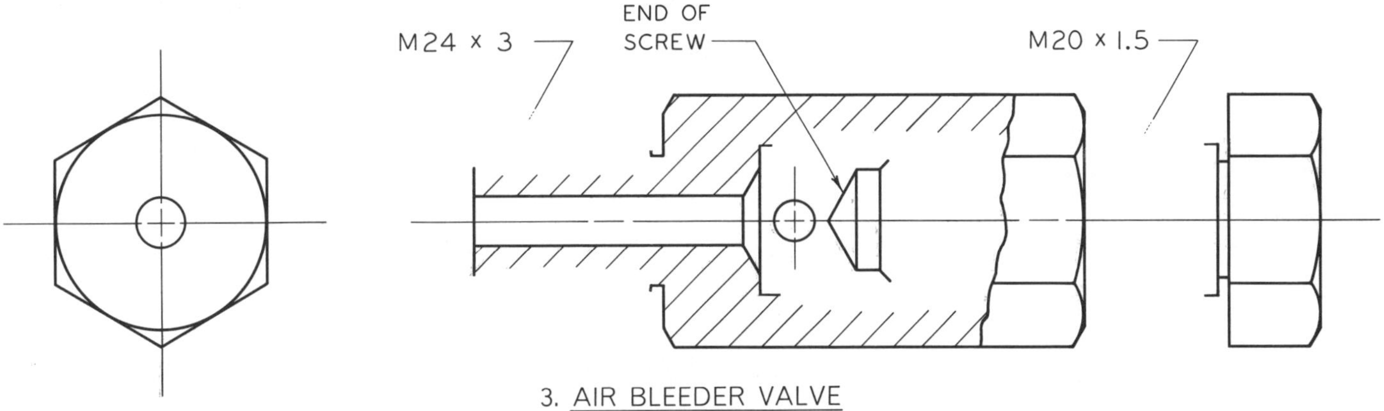
M24 × 3
END OF
SCREW
M20 × 1.5

3. AIR BLEEDER VALVE
Complete the views.

Bolt head on this end.

Nut and washer on this end.

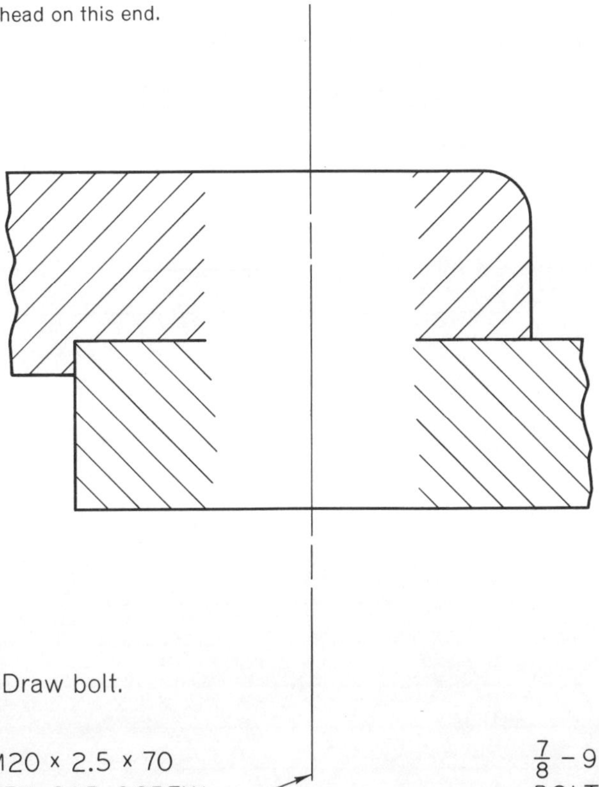

4. Draw bolt.

M20 × 2.5 × 70
HEX CAP SCREW
& HEX NUT

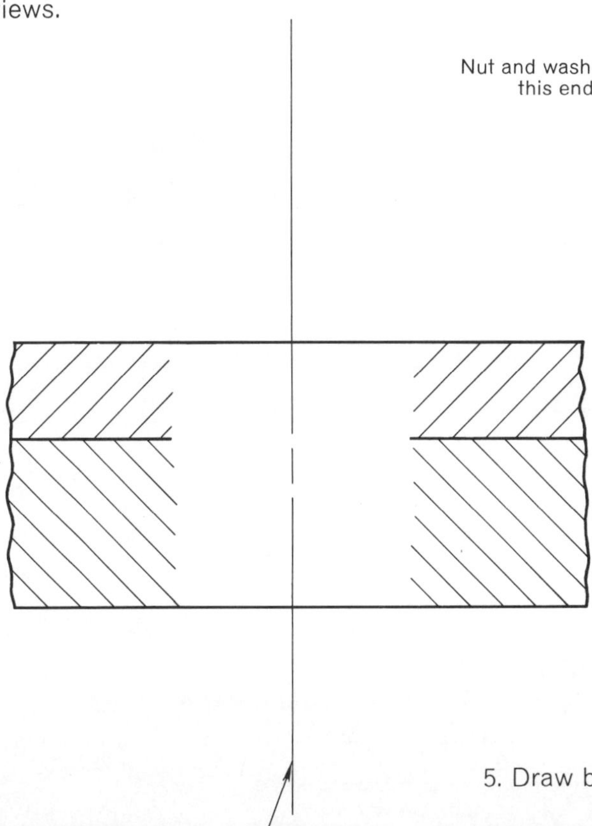

5. Draw bolt.

$\frac{7}{8}$ – 9 UNC × 2$\frac{3}{4}$ SQ HD
BOLT & NUT WITH REG
AMER NATL STD LOCK WASH.

DRAWN BY

FILE NO.

DRAW

(1) Complete the views of the nonintersecting tubes including correct visibility.

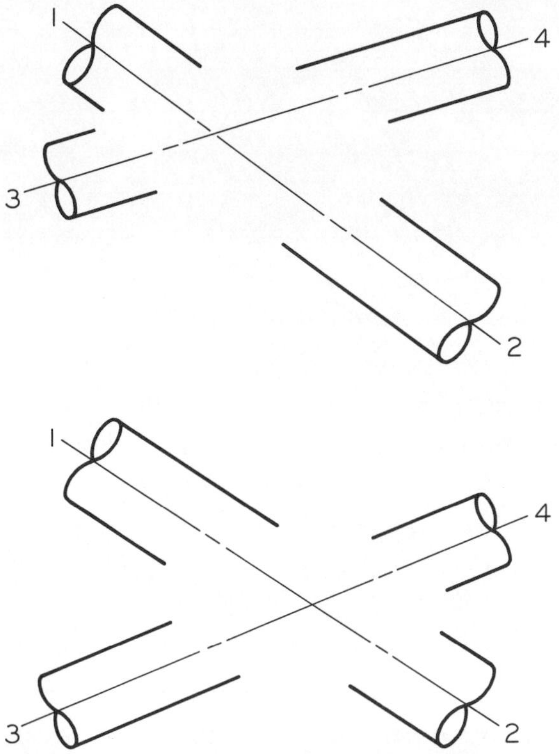
1
4
3
2
1
4
3
2

(2) Complete the views of the nonintersecting rods including correct visibility.

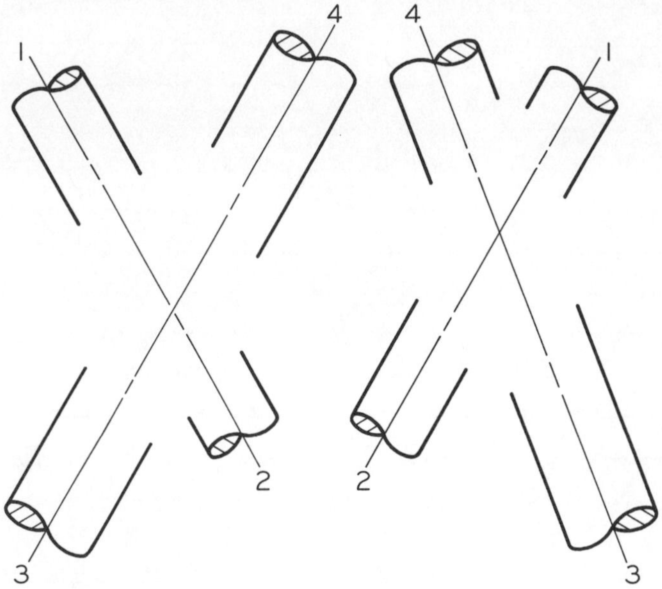
4
4
1
1
2
2
3
3

(3) Points 1, 2, 3, and 4 are the vertices of a tetrahedron. Complete the views and add a left-side view, all with correct visibility.

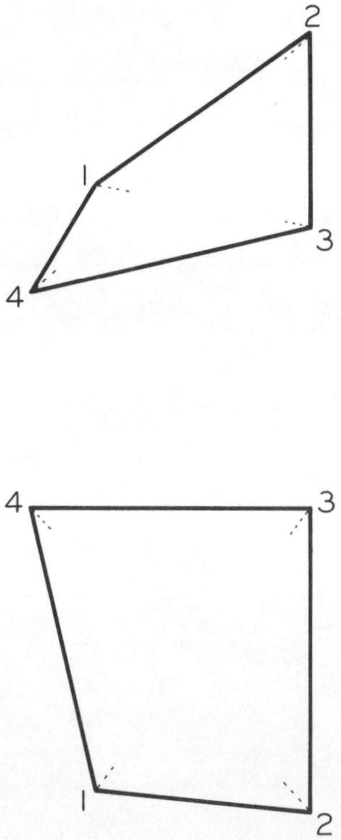
2
1
3
4
4
3
1
2

(4) Parallelogram 1-2-3-4 is the base of a pyramid. Point V is the vertex. Complete the views and add a right-side view.

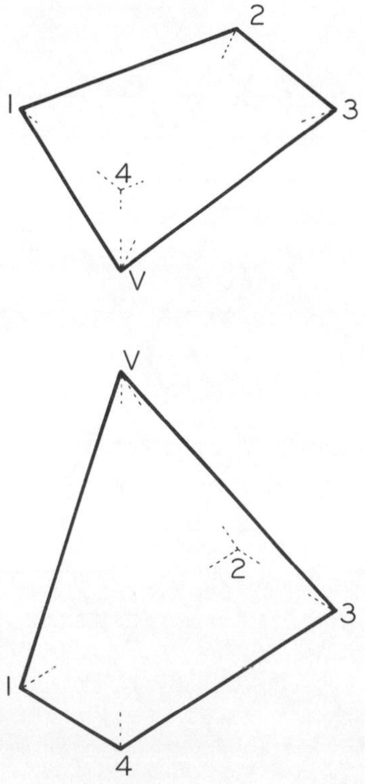
2
1
3
4
V
V
2
3
1
4

VISIBILITY POINTS AND LINES	DRAWN BY	FILE NO.	DRAWING 48

	DRAWN BY	FILE NO.	DRAWI

(1) Locate the front and side views of line 1-2. Add the top view.

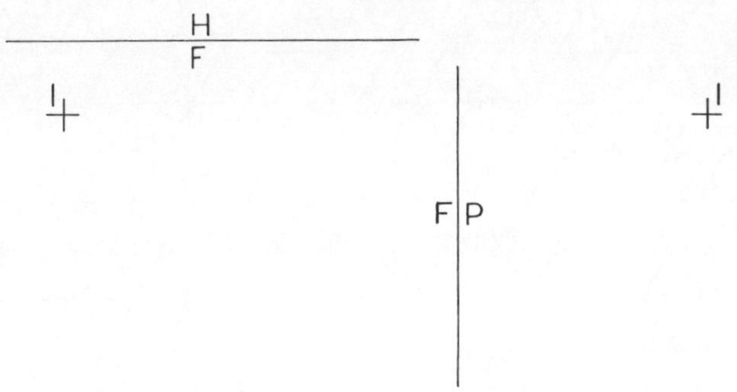
H
F
1
1
F P

(2) Find the true length of line 2-3.

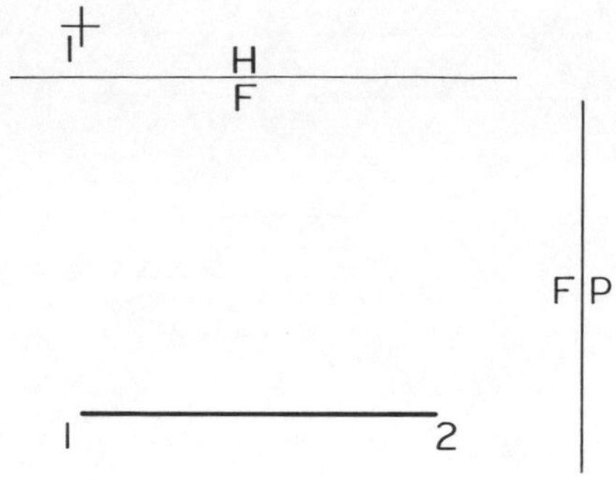
1
H
F
F P
1
2

(3) Complete the three views of triangle 1-2-3.

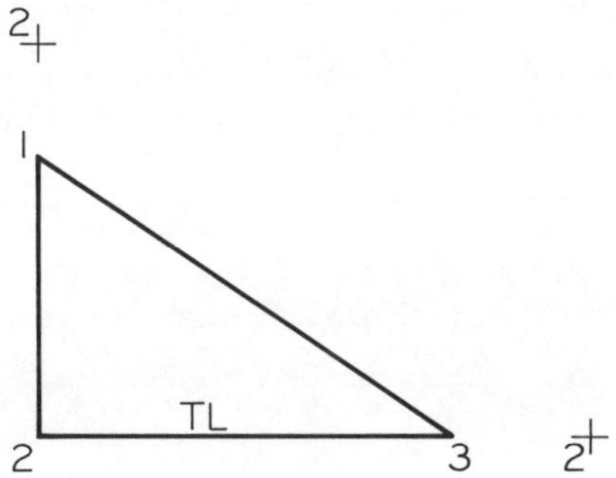
2
1
TL
2
3
2

(4) Find the true length of connecting line 5-6.

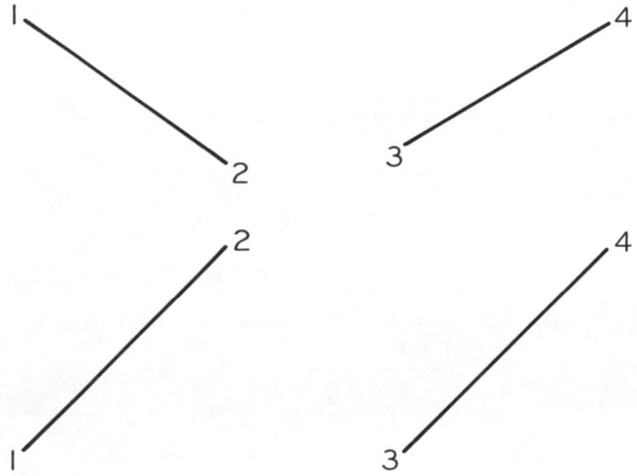
1
2
4
3
2
1
4
3

(5) Move point 4 vertically in space to a new position 4′ such that line 3-4′ intersects line 1-2.

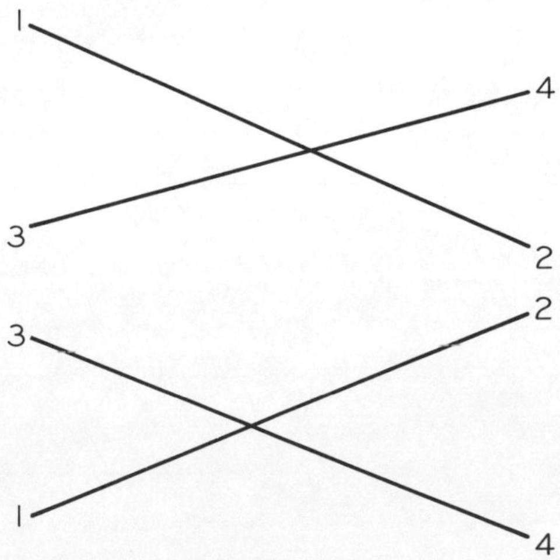
1
4
3
2
2
3
1
4

(6) Complete the front and top views of pyramid V-1-2-3.

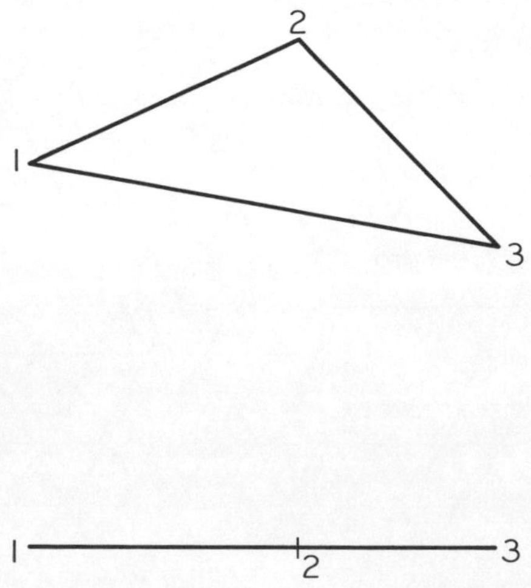
2
1
3
1
2
3

	DRAWN BY	FILE NO.	DRAW

(1) Find and measure the true length of line 1-2 and the angle it forms with horizontal (∠H).

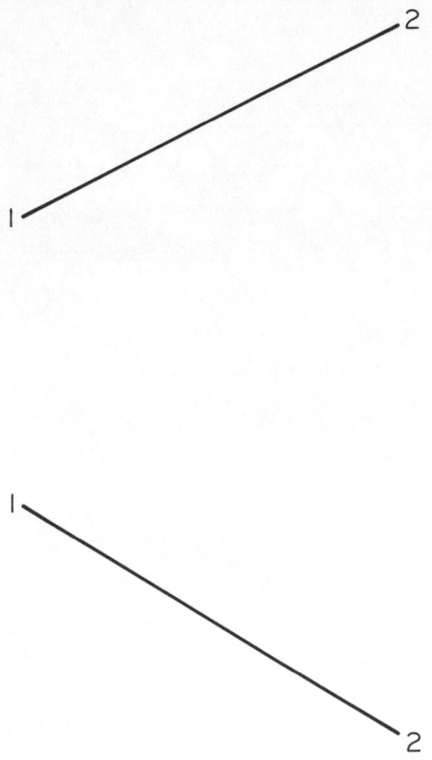

2
1
1
2

(2) Find and measure the true length of pipe center line 1-2 and the angle it forms with the frontal wall (∠F). Scale: 1/50.

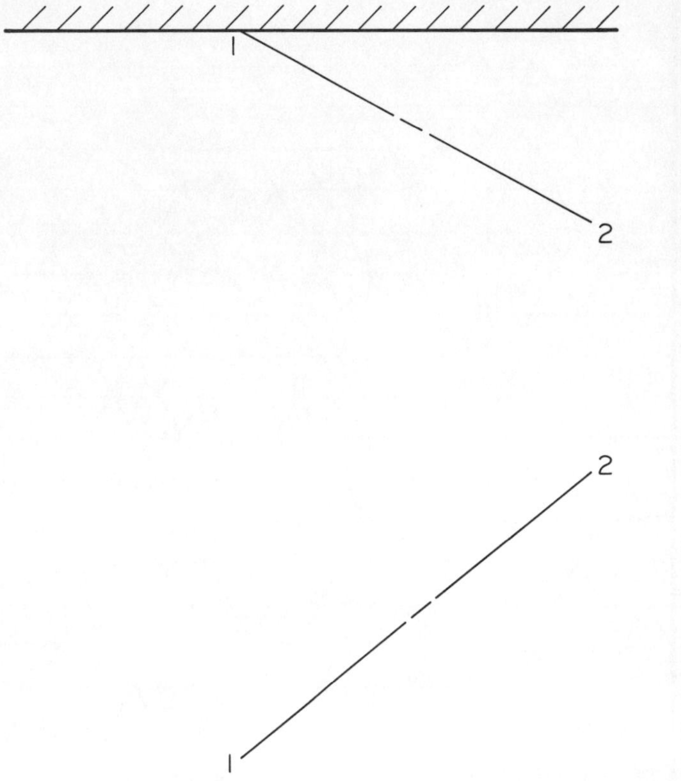

1
2
2
1

(3) Find and measure the true length of line 1-2 and the angle it forms with a profile plane (∠P).

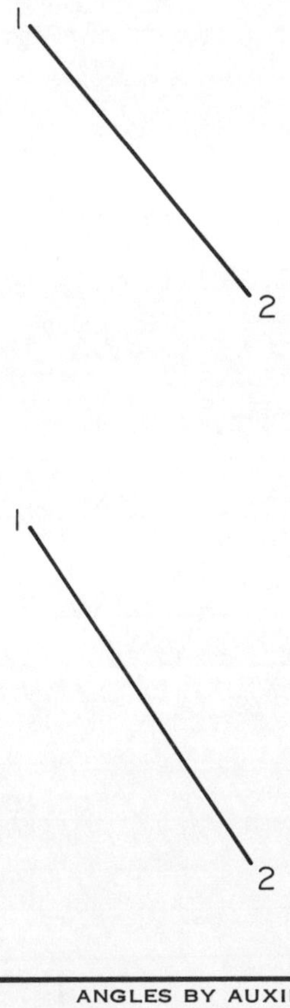

1
2
1
2

(4) Center line 1-2 of a ramp slopes downward from point 1 at 20° and is 116 m long. Complete the views. Scale: 1/2500.

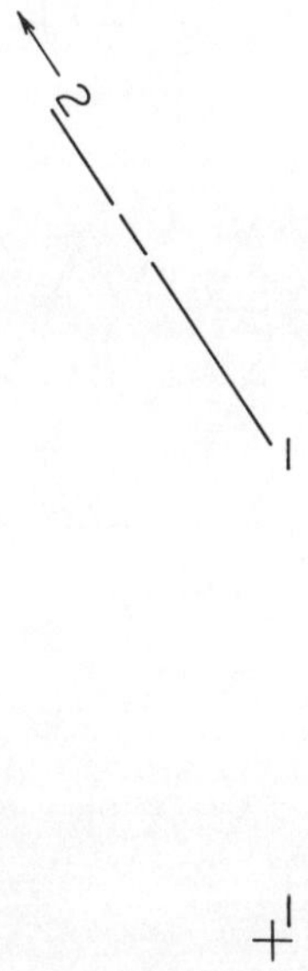

2
1
1

ANGLES BY AUXILIARY VIEWS TRUE LENGTH OF LINE	DRAWN BY	FILE NO.	DRAWING 50

DRAWN BY

FILE NO.

DRAWI

(1) Measure the length, bearing, and grade of tunnel 1-2. Scale: 1/5000.

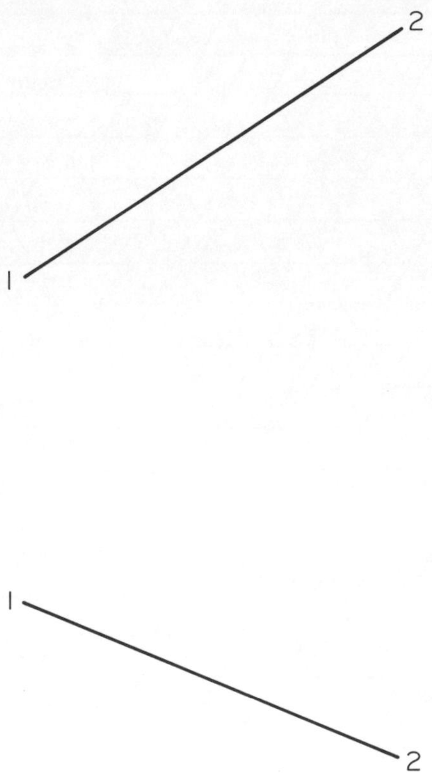
2
1
1
2

(2) Measure the length, bearing, and grade of sluice 1-2. Scale: 1/750.

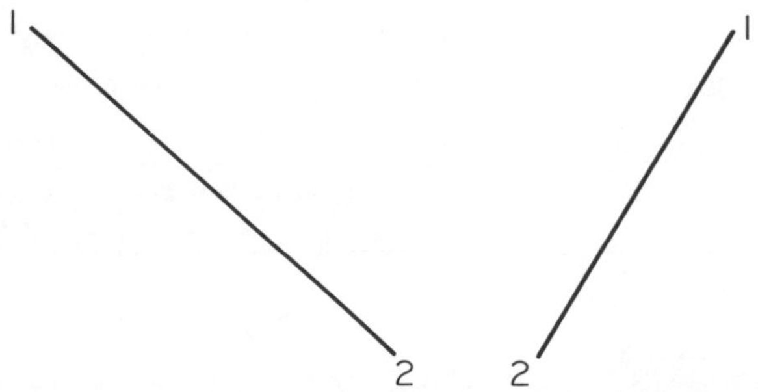
1
2
2
1

(3) A tunnel bears N40°E from point 1 at a grade of −30%, to a point 2 that is 214 m along the tunnel. Complete the views. Scale: 1/5000.

1
1

(4) If cable 3-2 has the same grade as cable 1-2, complete the front view. Measure the grade.

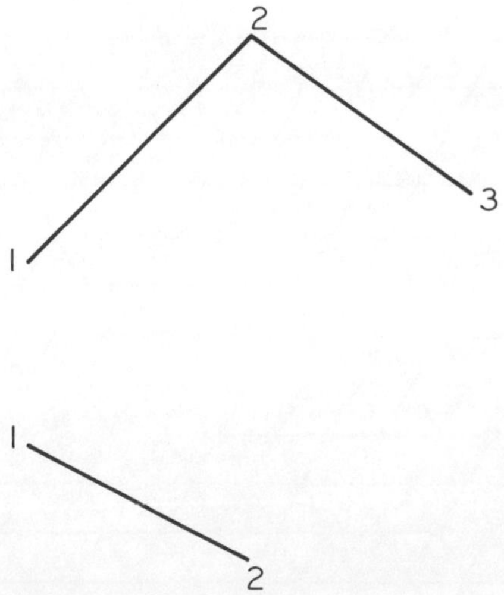
2
3
1
1
2

BEARING AND GRADE TRUE LENGTH OF LINE	DRAWN BY	FILE NO.	DRAWING 51

	DRAWN BY	FILE NO.	DRAWIN

(1) By revolution, find the true length and slope of line 1-2. Measure the bearing.

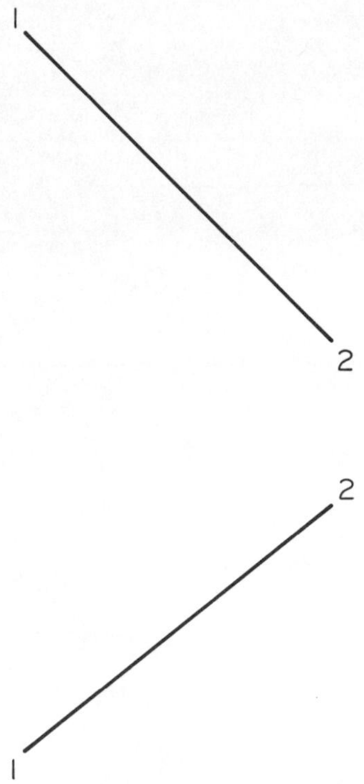
1
2
2
1

(2) Find the angles formed by line 1-2 and planes F and P. Measure the true length of line 1-2.

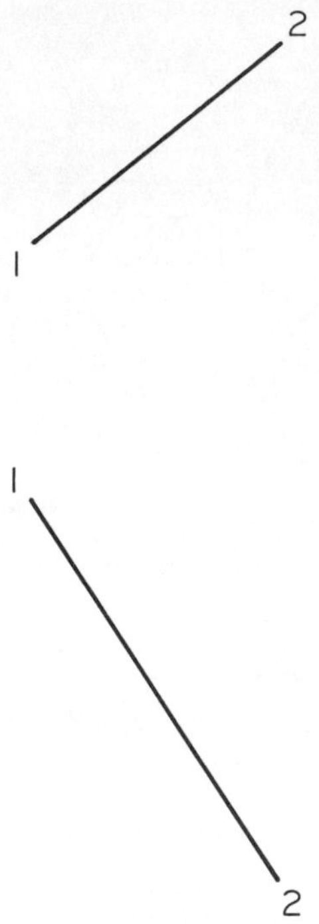
2
1
1
2

(3) Line 1-2 forms an angle of 30° with a frontal plane. Complete its front view.

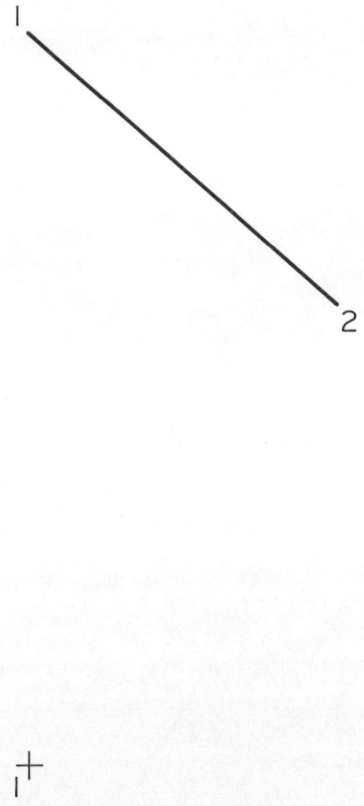
1
2
1

(4) An aircraft at point 1, on an azimuth course of N220°, is losing altitude at the rate of 400 m in 1000 m (map distance). Find the front and top views of a segment of the flight path.

1
1

REVOLUTION TRUE LENGTH OF LINE	DRAWN BY	FILE NO.	DRAWING 52

	DRAWN BY	FILE NO.	DRAWIN

(1) Obtain a point view of line 1-2.

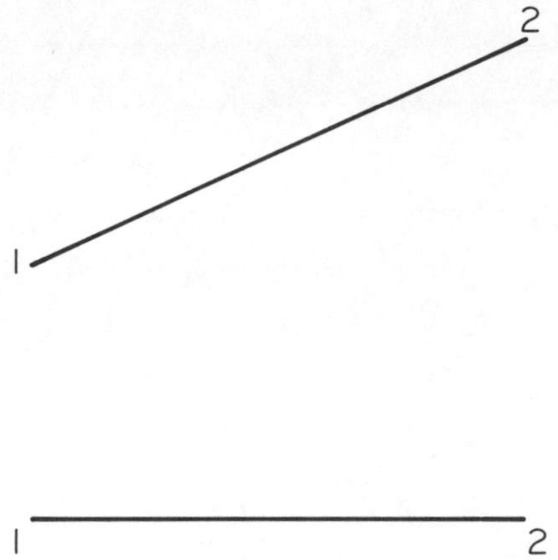

2
1
1
2

(2) Determine the clearance (minimum distance) between cylinder 1-2 and spherical tank 3. Scale: 1/250.

2
1
3
3
1
2

(3) Is point 1 nearer to line 2-3 or to line 4-5? Measure the true distances.

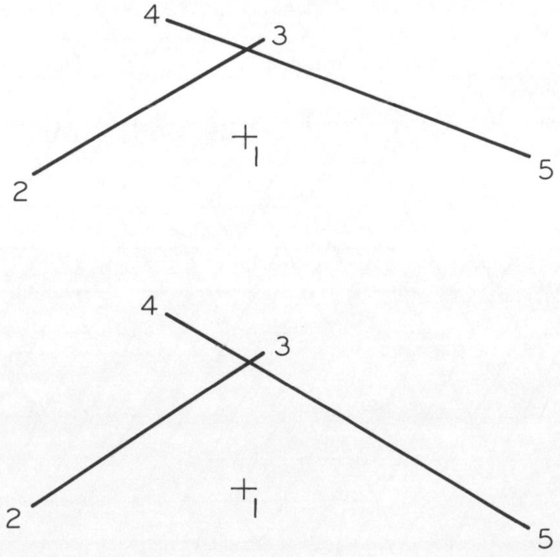

4
3
1
2
5
4
3
1
2
5

POINT VIEW TRUE LENGTH OF LINE	DRAWN BY	FILE NO.	DRAWING 53

	DRAWN BY	FILE NO.	DRAWIN

① Which, if either, of points 4 or 5 lies in plane 1-2-3?

ANS. ____________

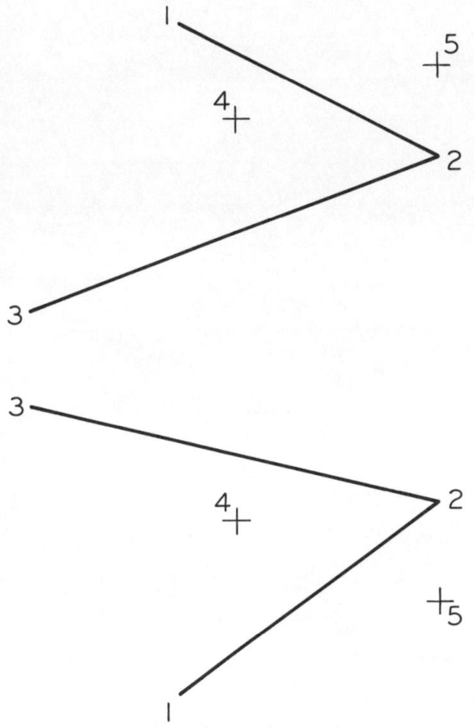
1
5
4
2
3
3
2
4
5
1

② Line 4-5 lies in plane 1-2-3. Complete the views of line 4-5.

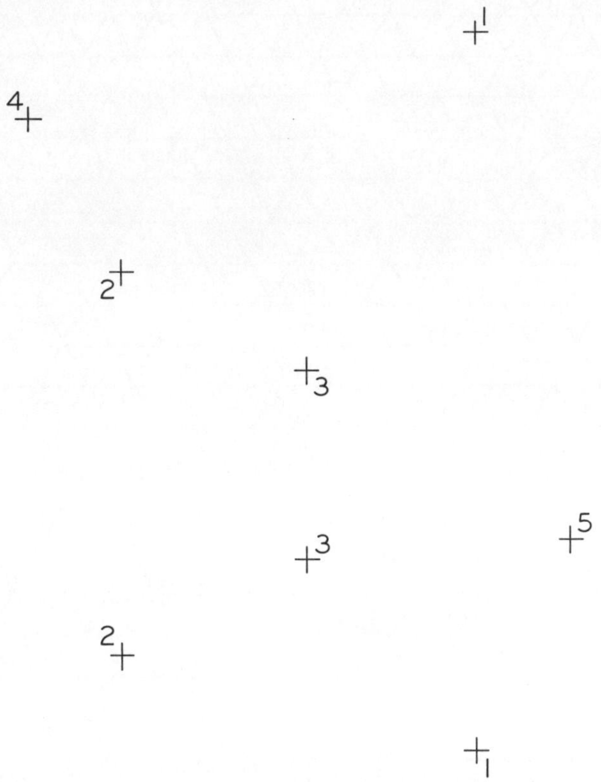
1
4
2
3
3
5
2
1

③ Quadrilateral 1-2-3-4 lies in plane 5-6-7. Find its front view.

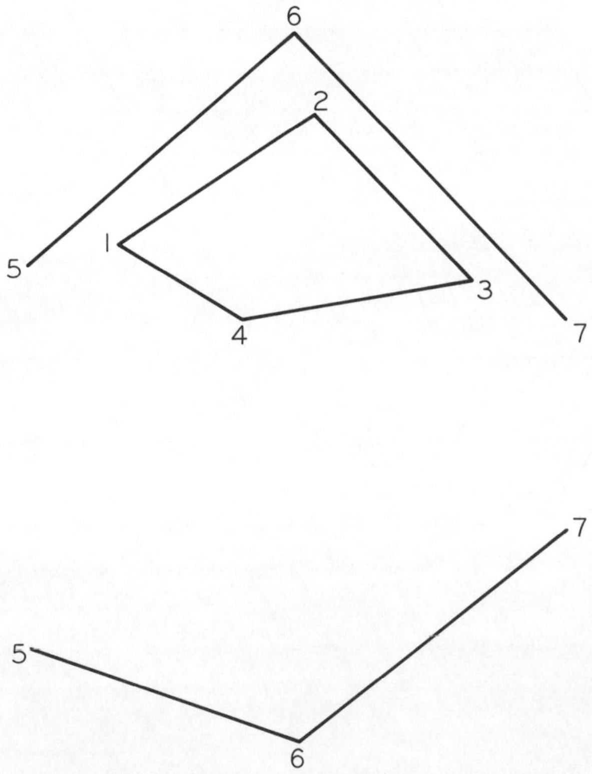
6
2
1
5
3
4
7
7
5
6

④ Find a point P in plane 1-2-3 which lies 16mm below point 1 and 10mm in front of point 3.

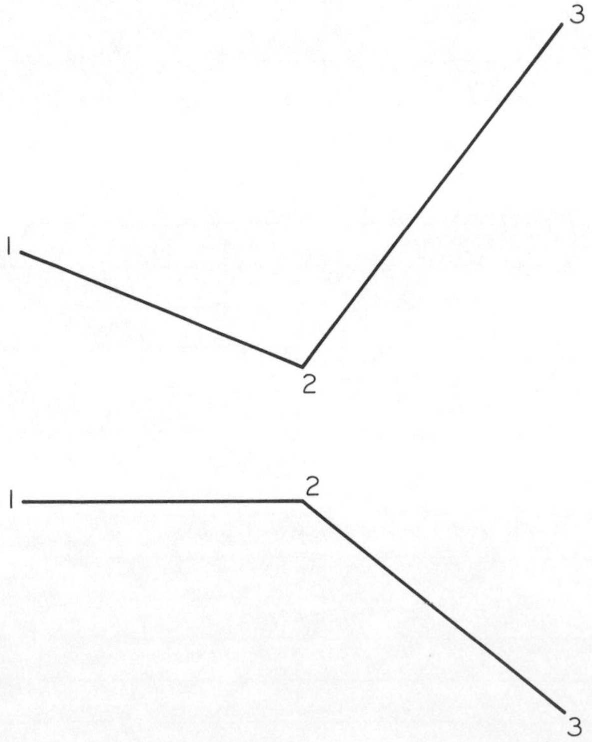
3
1
2
2
1
3

	DRAWN BY	FILE NO.	DRAWIN

(1) Construct a view showing the true size of triangle 1-2-3.
Calculate the area of the triangle. Scale: 1/2500.

AREA = ____________

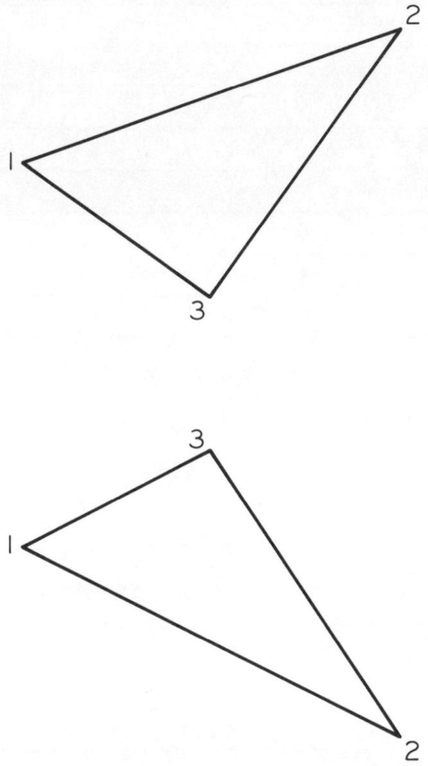

2
1
3
3
1
2

METRIC

(2) Pipe lines 1-2 and 3-4 are connected with a feeder branch using 45° lateral fittings. One fitting is located at point 5 on line 3-4.
Find the views of the feeder branch.

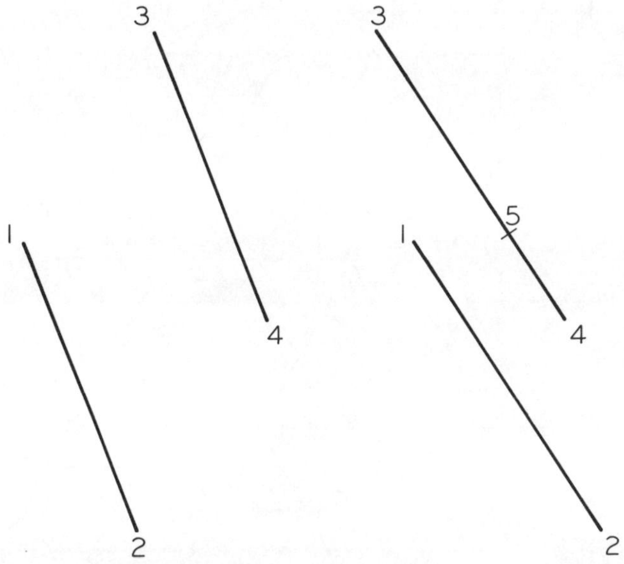

3
3
5
1
1
4
4
2
2

TRUE SIZE	DRAWN BY	FILE NO.	DRAWING
PLANES			55

	DRAWN BY	FILE NO.	DRAWI

(1) Find the top and front views of center O and the points of tangency of a circle inscribed in triangle 1-2-3.

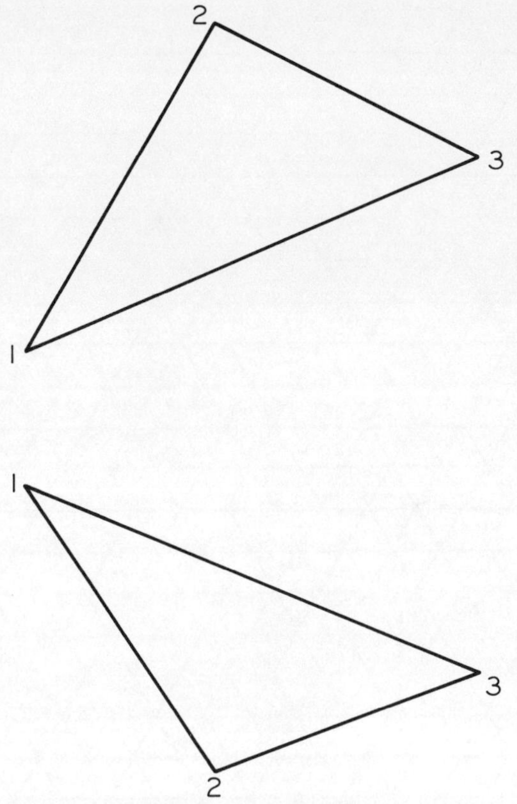

2
3
1
1
3
2

(2) Structural member 1-2 is connected to point 3 with another structural member whose length is 4.5 m. Find the front and side views of the connecting member. Scale: 1/100.

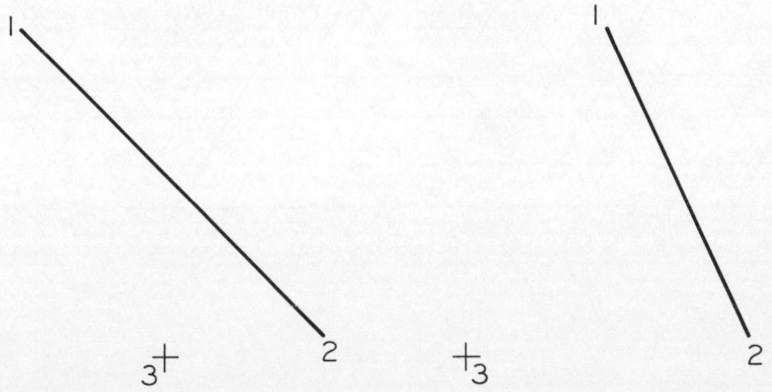

1
2
3
1
3
2

TRUE SIZE PLANES	DRAWN BY	FILE NO.	DRAWING 56

	DRAWN BY	FILE NO.	DRAWI

(1) Find the piercing point of line 1-2 in plane 3-4-5.

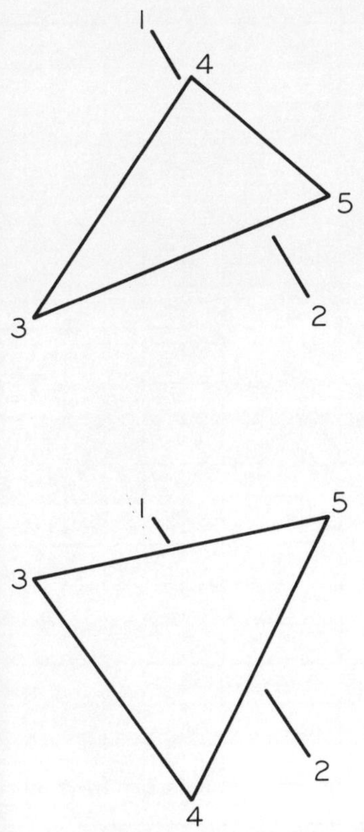
1
4
5
3
2
1
5
3
2
4

(2) Establish the piercing point of laser beam 1-2 in plane 3-4-5-6.

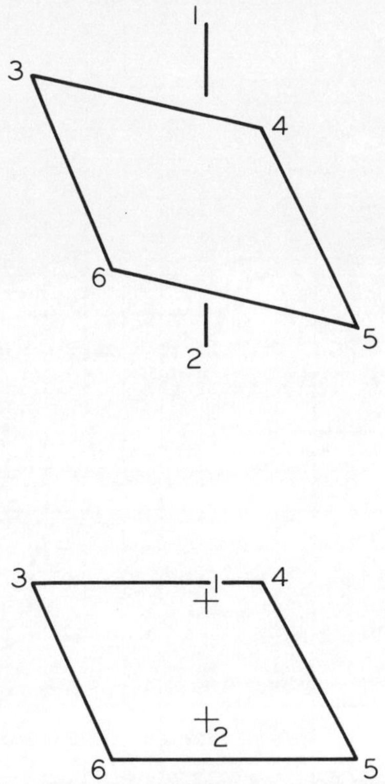
1
3
4
6
5
2
3
1
4
2
6
5

(3) Show the piercing points of line 1-2 in the surfaces of the pyramid.

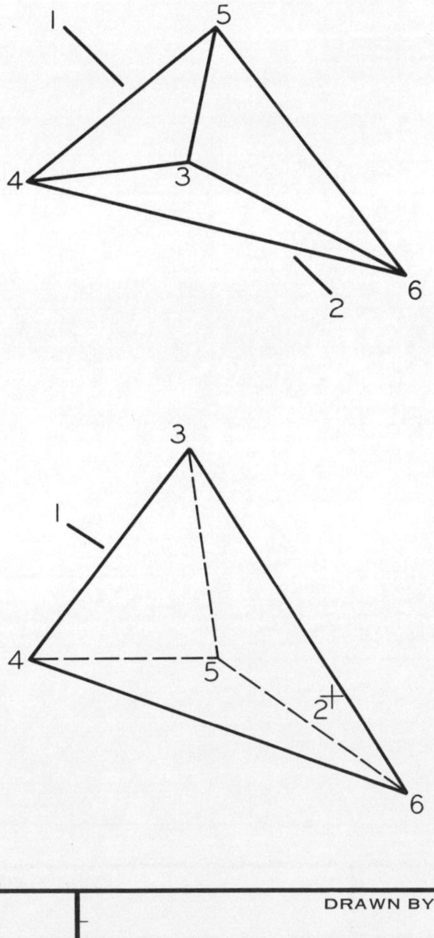
1
5
4
3
2
6
3
1
4
5
2
6

PIERCING POINTS—EDGE-VIEW METHOD LINES AND PLANES	DRAWN BY	FILE NO.	DRAWING 57

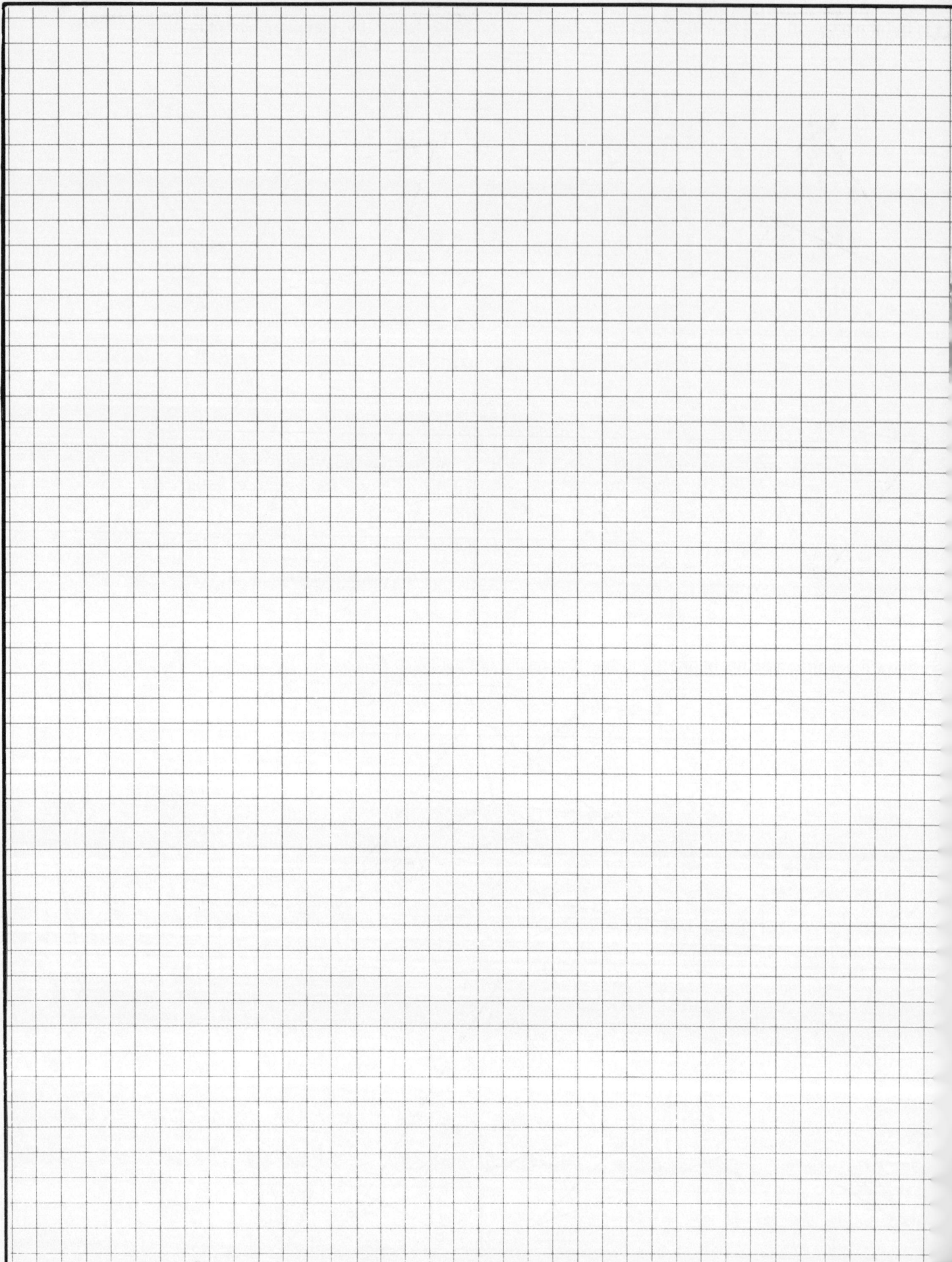

	DRAWN BY	FILE NO.	DRAWI

(1) Locate the piercing point of pipe 1-2 in plane 3-4-5.

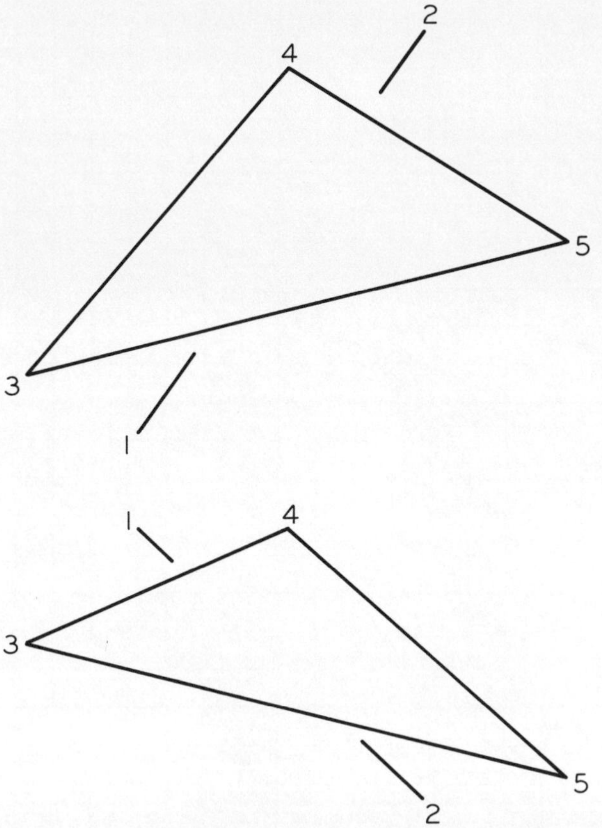
2
4
5
3
1
1
4
3
5
2

(2) Find the point at which light ray 1-2 strikes mirror 3-4-5-6.

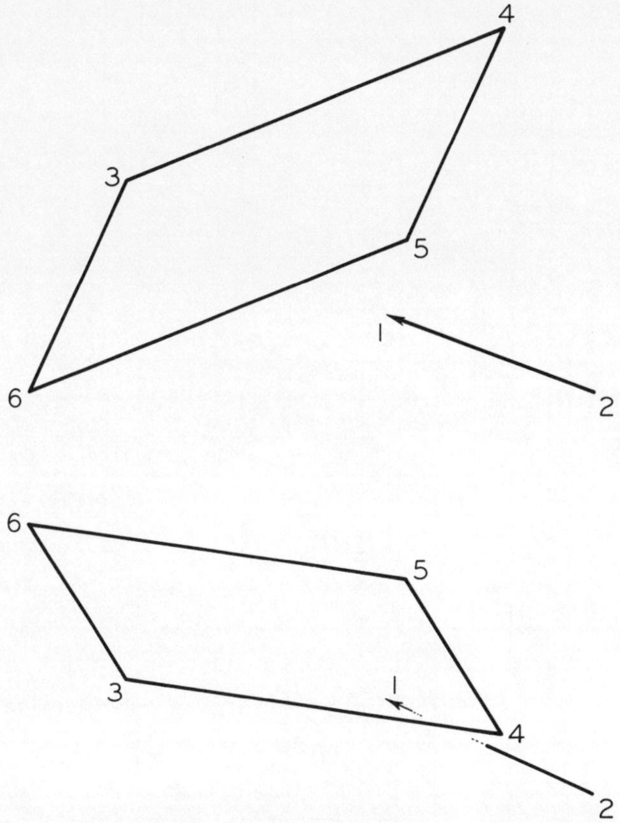
4
3
5
1
6
2
6
5
3
1
4
2

(3) Find the piercing points of line 1-2 and the pyramid.

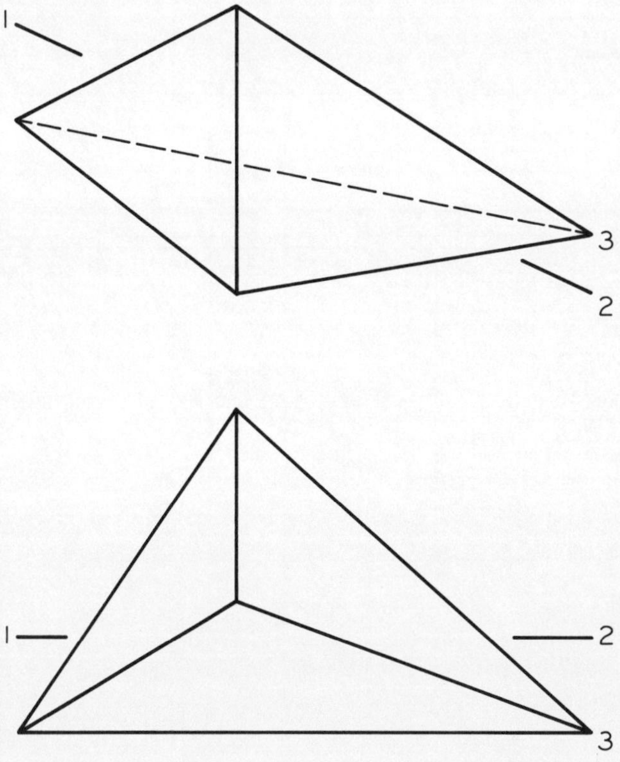
1
3
2
1
2
3

(4) Show the piercing point of shaft 1-2 in plane 3-4-5.

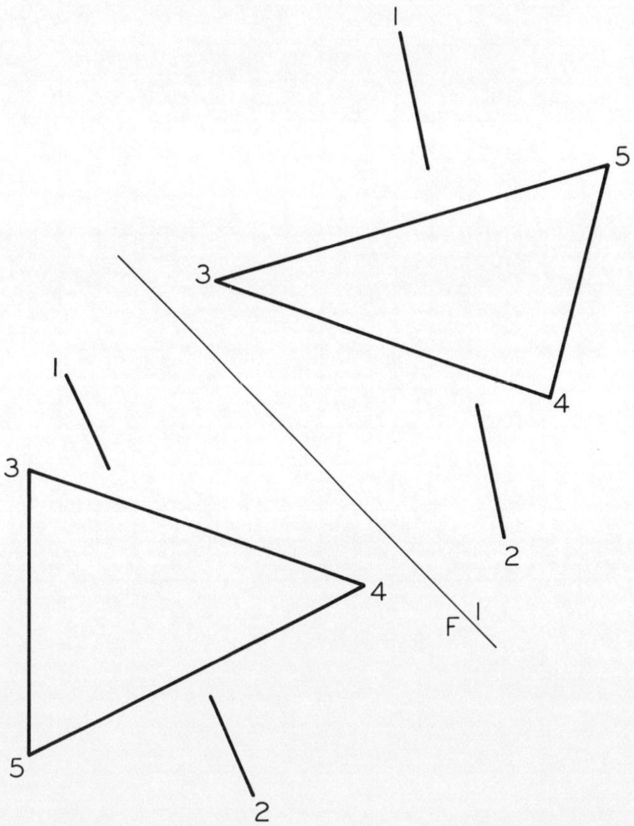
1
5
3
4
1
3
2
4
F
1
5
2

PIERCING POINTS—CUTTING-PLANE METHOD LINES AND PLANES	DRAWN BY	FILE NO.	DRAWING 58

	DRAWN BY	FILE NO.	DRAW

(1) By the edge-view method, find the line of intersection of the two planes.

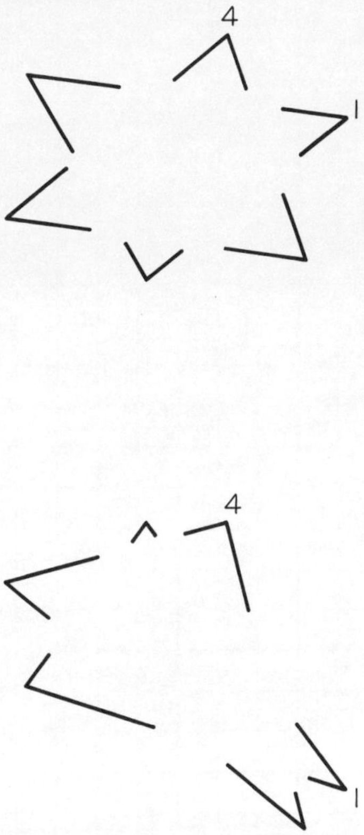
4
1
4
1

(2) By the cutting-plane method, show the line of intersection of the two surfaces.

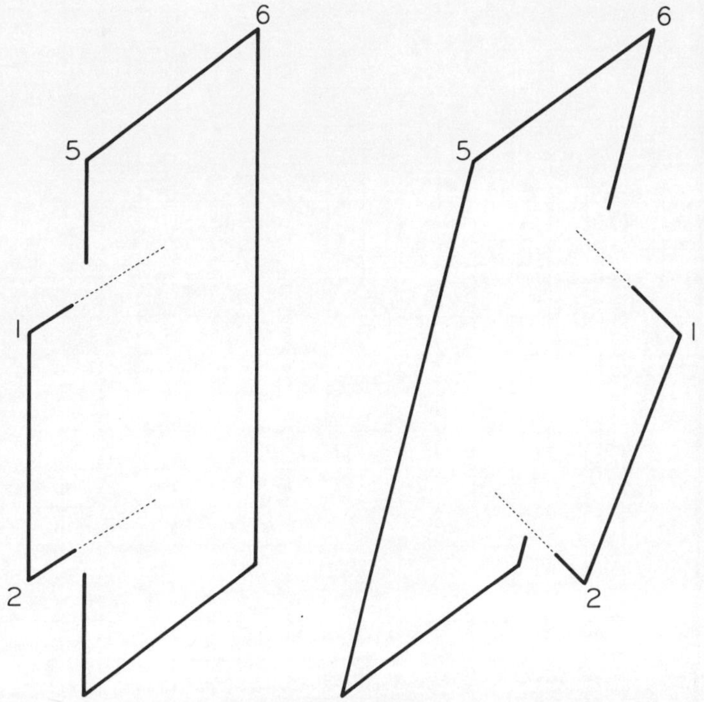
6
5
1
2
6
5
1
2

(3) By the cutting-plane method, establish the line of intersection of the two planes.

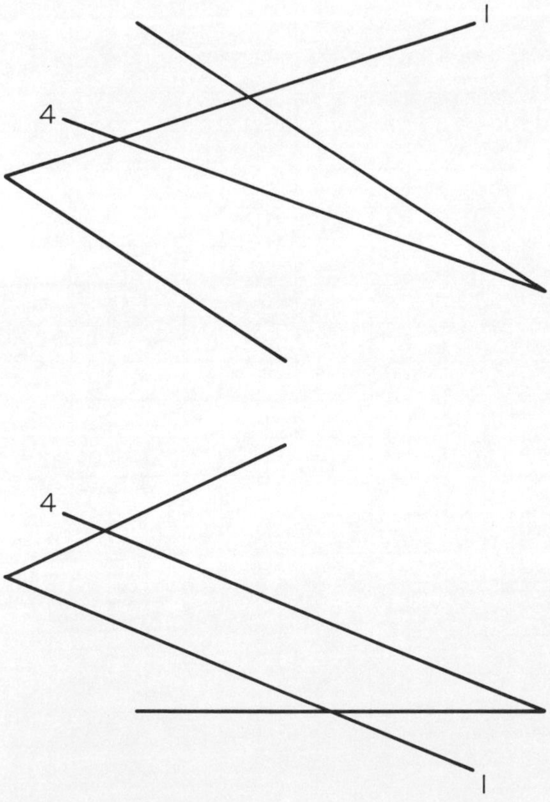
1
4
4
1

(4) By the special cutting-plane method, locate the intersection of the two planes.

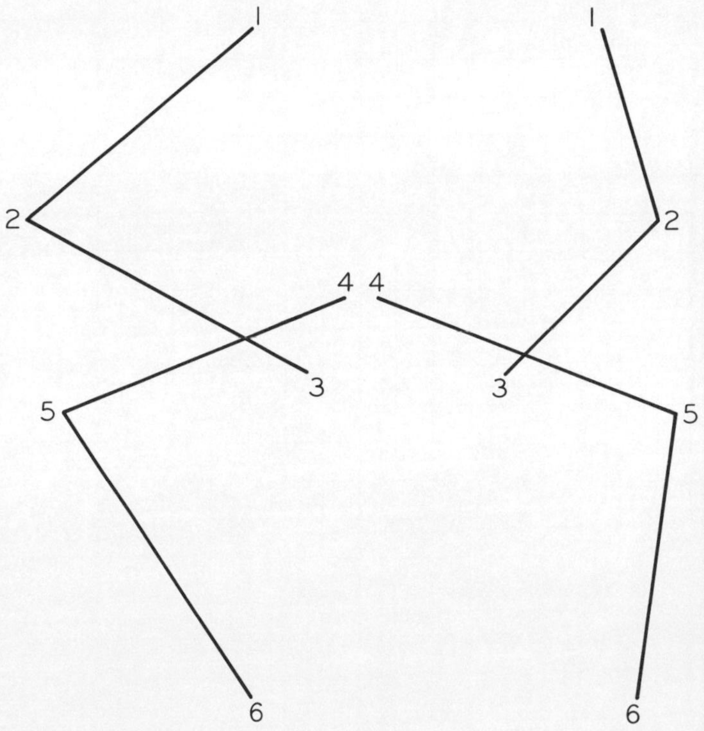
1
2
4
3
5
6
1
2
4
3
5
6

INTERSECTIONS PLANES	DRAWN BY	FILE NO.	DRAWING 59

DRAWN BY

FILE NO.

DRAWIN

(1) Find the dihedral angles between the lateral faces of the prism.

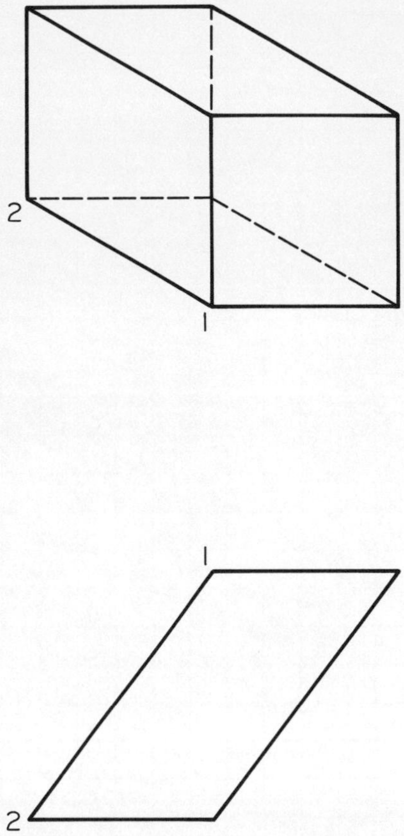
2
1
1
2

(2) Determine the angle between the roof planes.

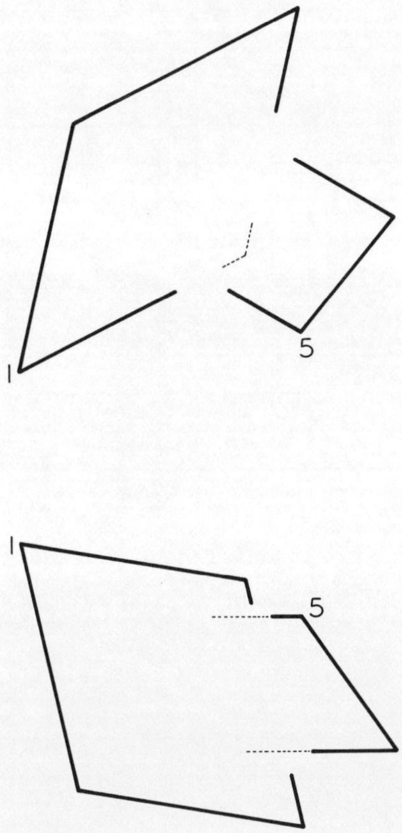
5
1
1
5

DIHEDRAL ANGLE PLANES	DRAWN BY	FILE NO.	DRAWING 60

DRAWN BY

FILE NO.

DRAWI

(1) Find the angle between control cable 1-2 and bulkhead 3-4-5-6.

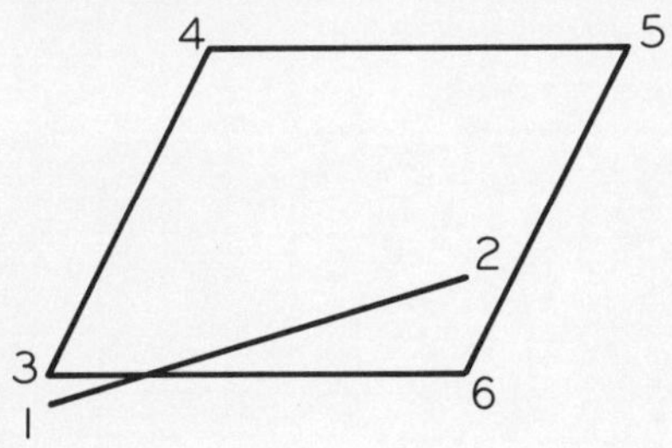
4
5
2
3
1
6

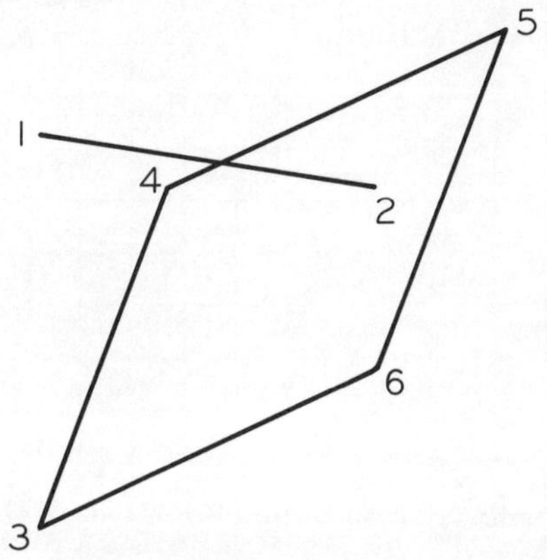
5
1
4
2
6
3

(2) Establish the views of 38 mm line 1-2 such that the line 1-2 forms an angle of 25° with the given surface.

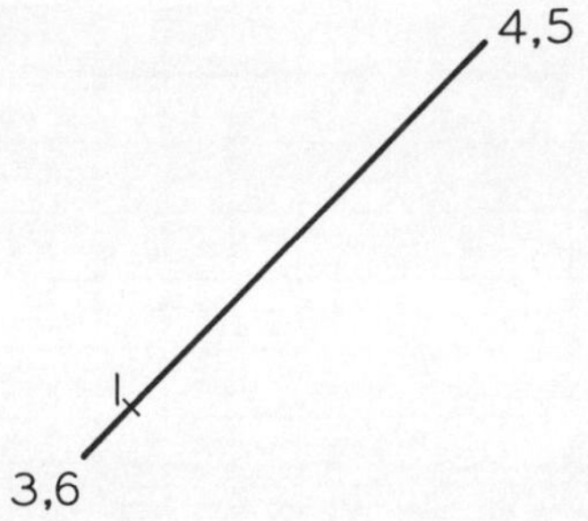
4,5
1
3,6

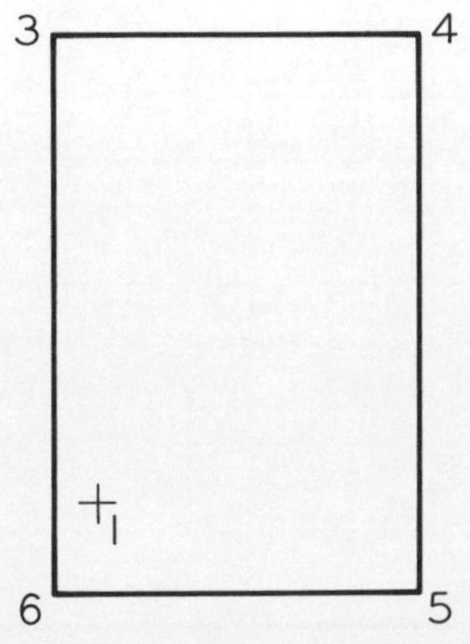
3
4
1
6
5

ANGLE BETWEEN LINE AND PLANE LINES AND PLANES	DRAWN BY	FILE NO.	DRAWING 61

	DRAWN BY	FILE NO.	DRAWIN

(1) Which lines are parallel?

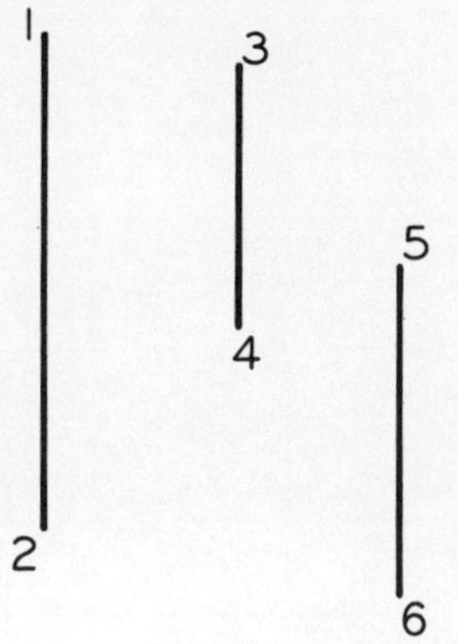

1
2
3
4
5
6

LINES
1-2 ______
3-4 ______
5-6 ______

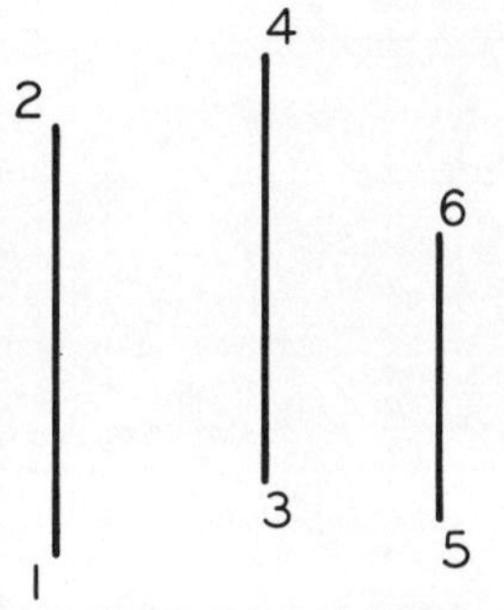

2
1
4
3
6
5

(2) By inspection only, determine which of the given lines 4-5, 6-7, or 8-9 is/are in plane 1-2-3.

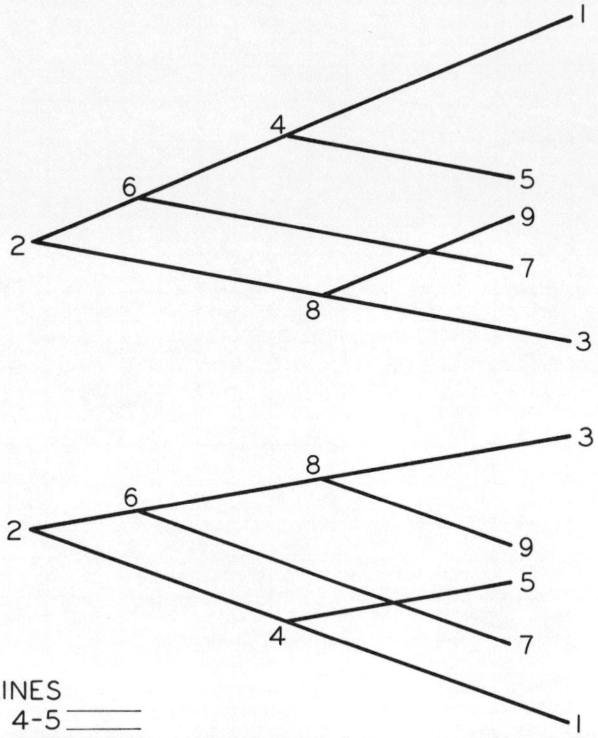

1
4
5
6
9
2
7
8
3
3
8
6
2
9
5
4
7
1

LINES
4-5 ______
6-7 ______
8-9 ______

(3) Complete the side view of plane 1-2-3, which is parallel to line 4-5.

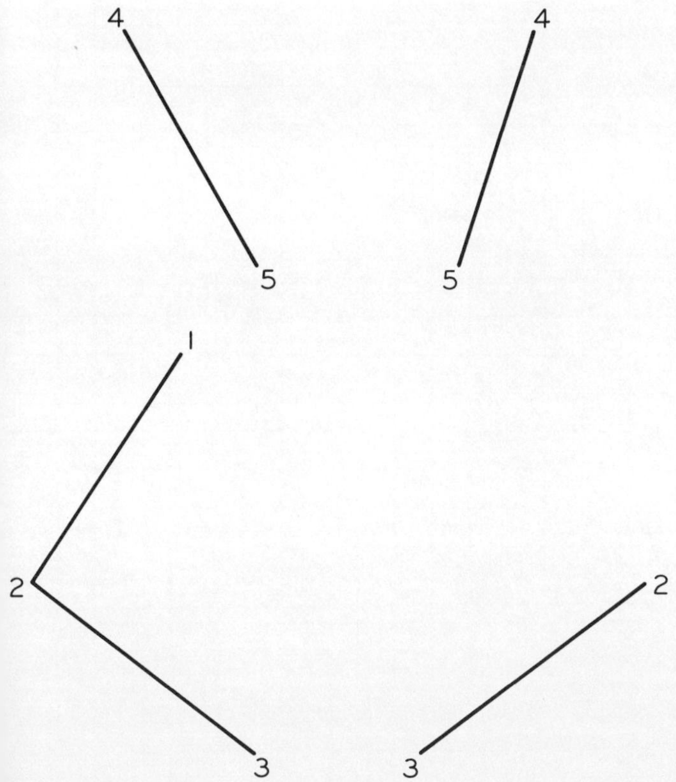

4
5
4
5
1
2
3
2
3

(4) Through point 7 draw line 7-8 parallel to planes 1-2-3 and 4-5-6.

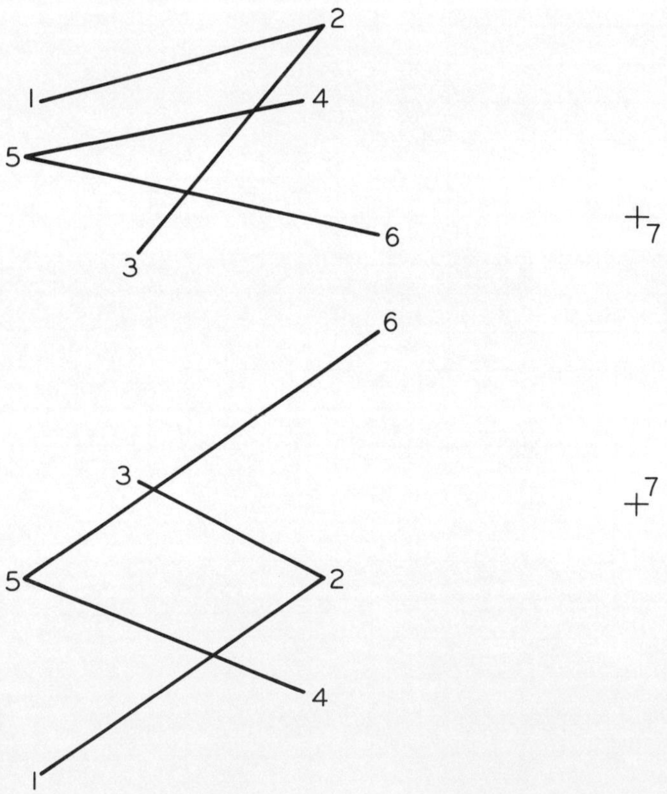

2
1
4
5
6
7
3
6
3
7
5
2
4
1

LINES AND PLANES PARALLELISM	DRAWN BY	FILE NO.	DRAWING 62

	DRAWN BY	FILE NO.	DRAWI

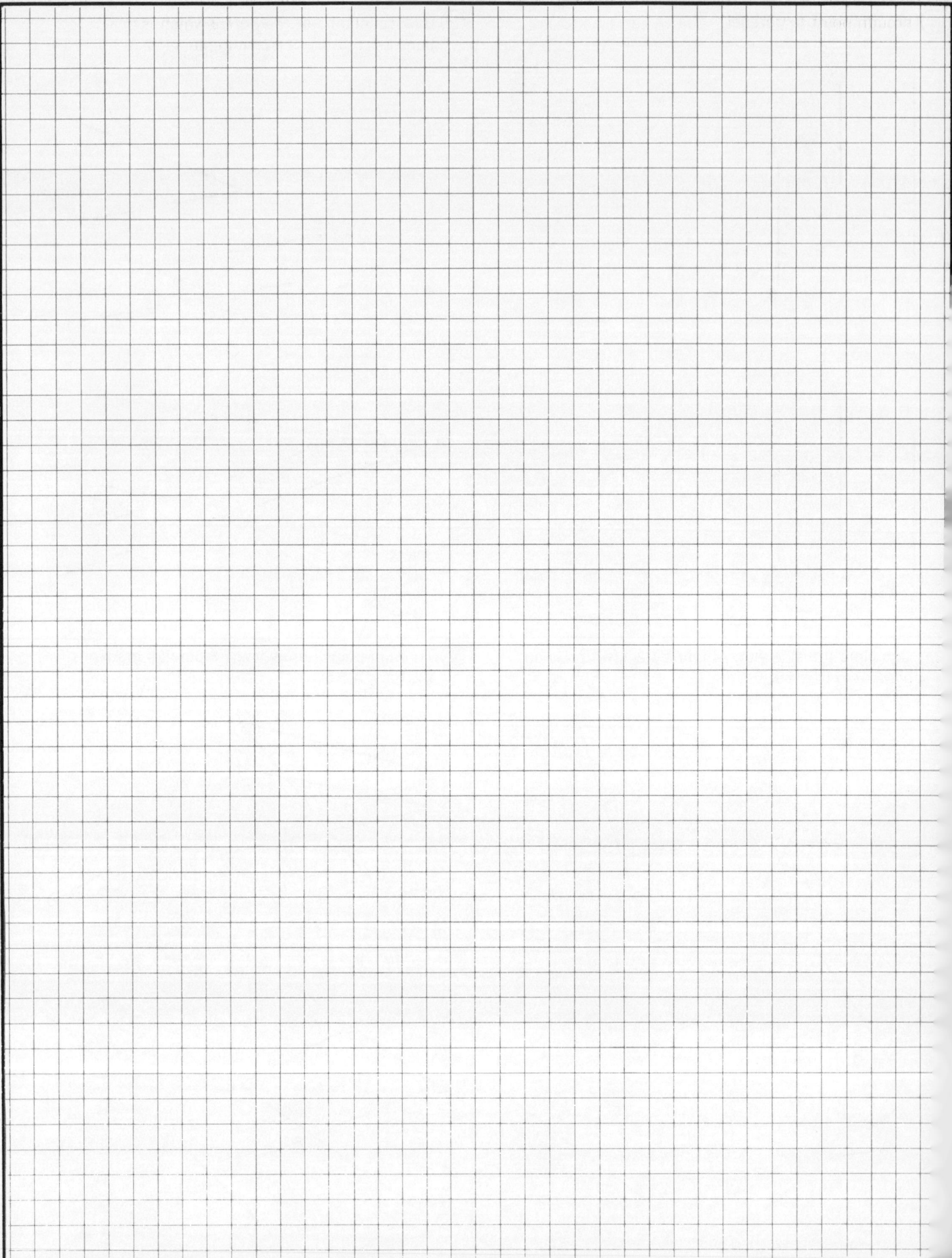

(1) Through point 1 draw line 1-2 which has a bearing of N30°E and is parallel to surface 3-4-5.

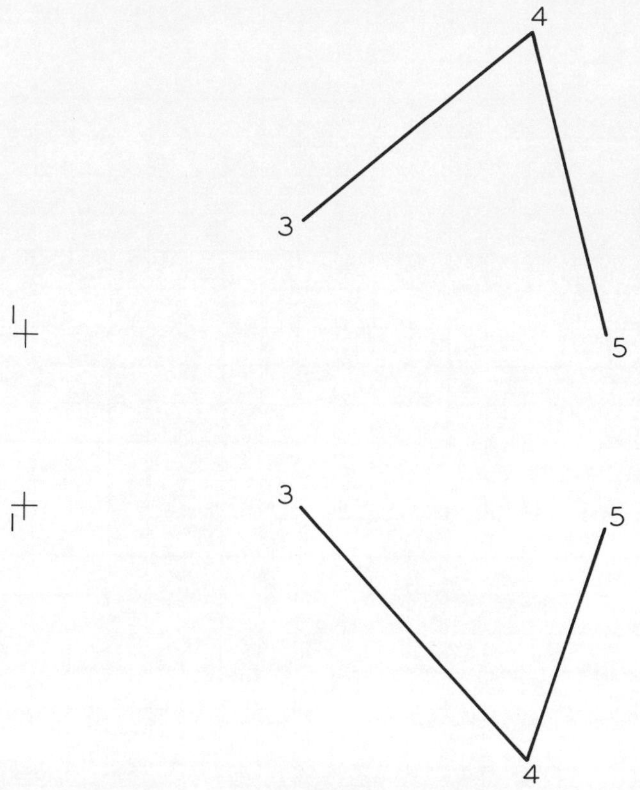
4
3
1
5
1
3
5
4

(2) Through point 1 draw line 1-2 which is parallel to plane 3-4-5 and intersects line 6-7.

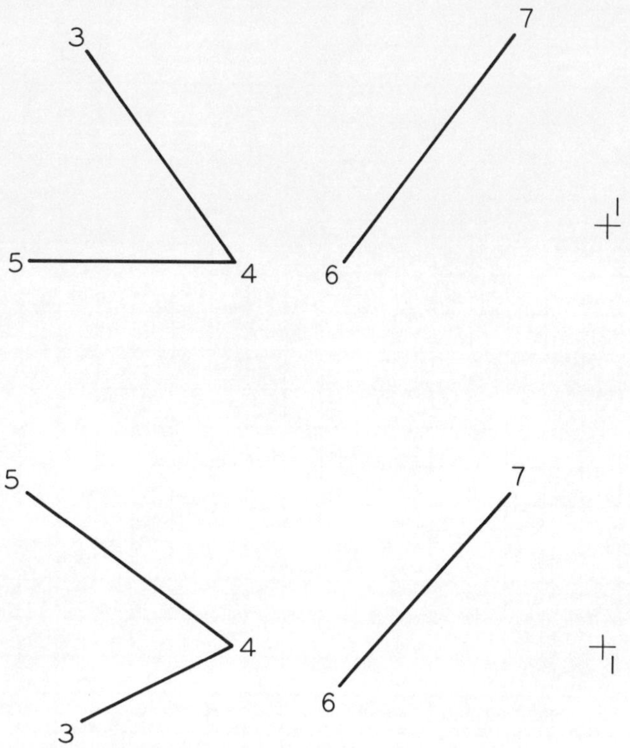
3
7
1
5
4
6
5
7
4
1
6
3

(3) Establish pipe 1-2 which is parallel to pipe 3-4 and connects with pipes 5-6 and 7-8.

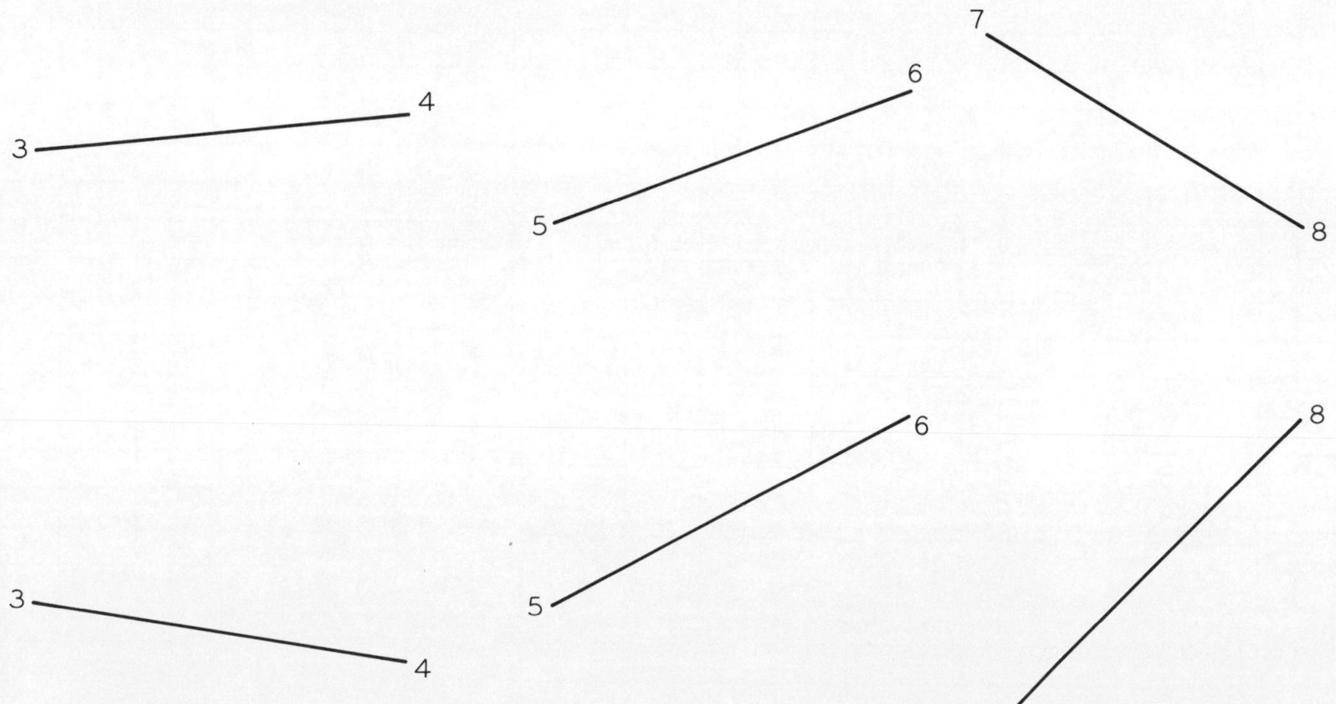
7
6
4
3
5
8
6
8
3
5
4
7

LINES AND PLANES PARALLELISM	DRAWN BY	FILE NO.	DRAWING 63

	DRAWN BY	FILE NO.	DRAW

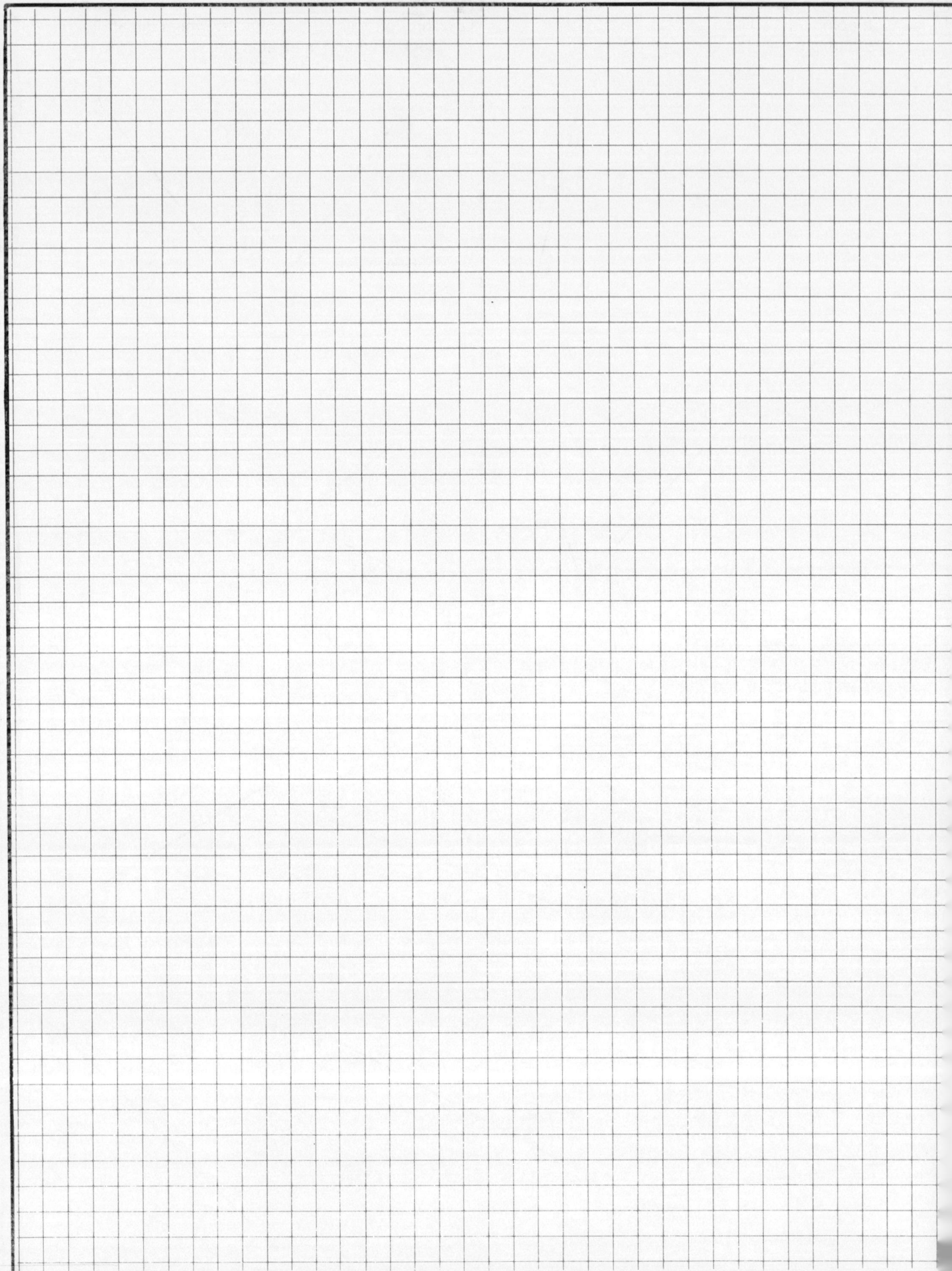

(1) Line 1-2 is perpendicular to line 3-4. Is point 2 on line 3-4? Complete the front view of line 1-2.

YES ________
NO ________

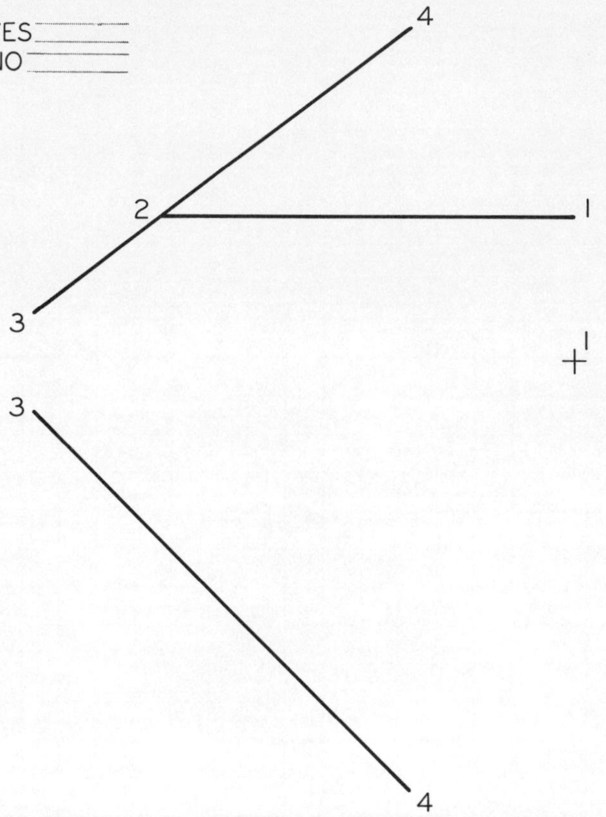
4
2
1
3
1
3
4

(2) Line 1-2 is perpendicular to line 3-4 and 82 mm in length. Complete the top view of the lines.

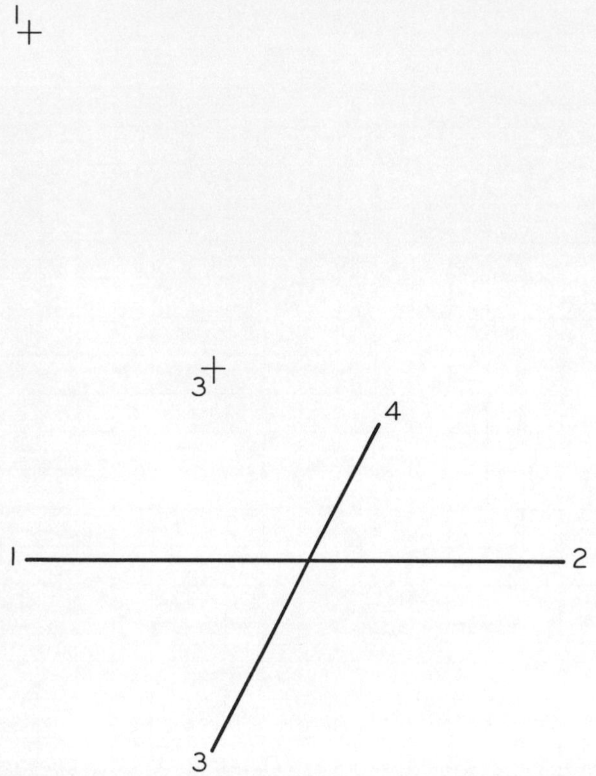
1
3
4
1
2
3

(3) Find the views of line 1-2 which is perpendicular to and intersects line 3-4.

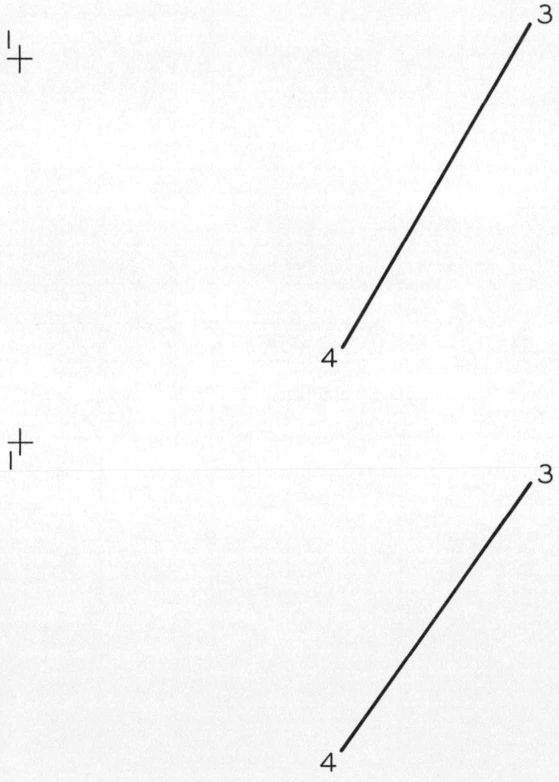
1
3
4
1
3
4

(4) Locate on line 1-2 the center O of the circle which passes through points 3 and 4.

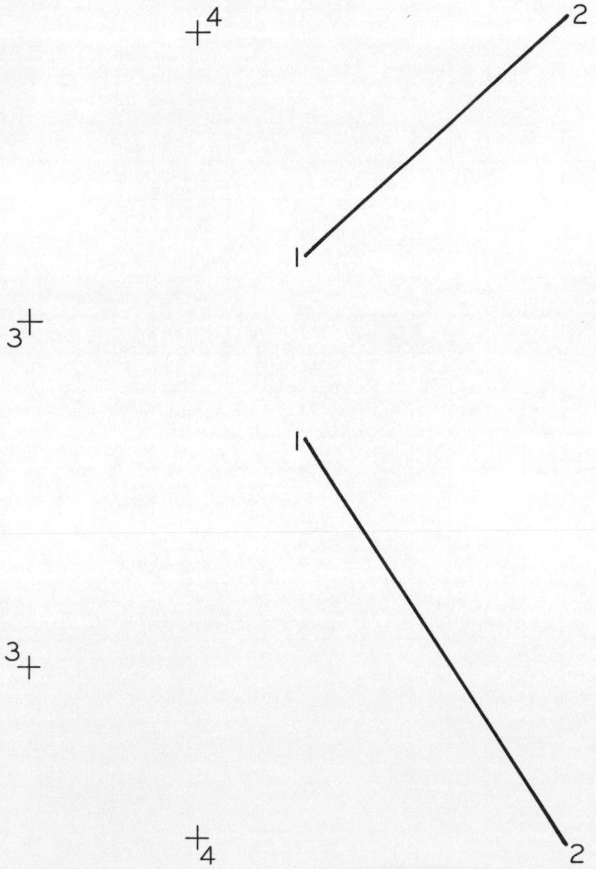
4
2
1
3
1
3
4
2

	DRAWN BY	FILE NO.	DRAWI

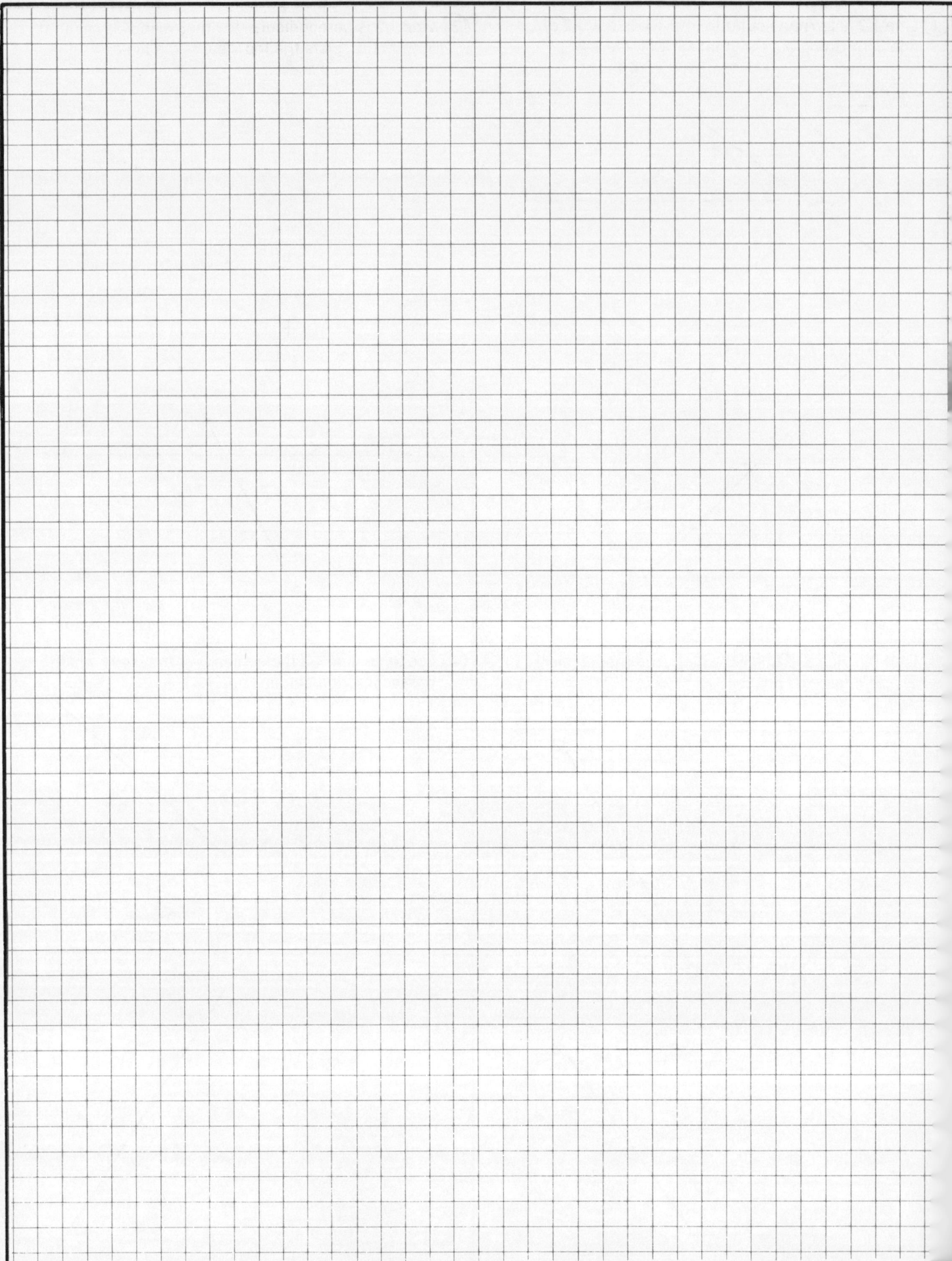

(1) Locate the shadow of point 1 on surface 2-3-4-5 if the light rays are perpendicular to the surface.

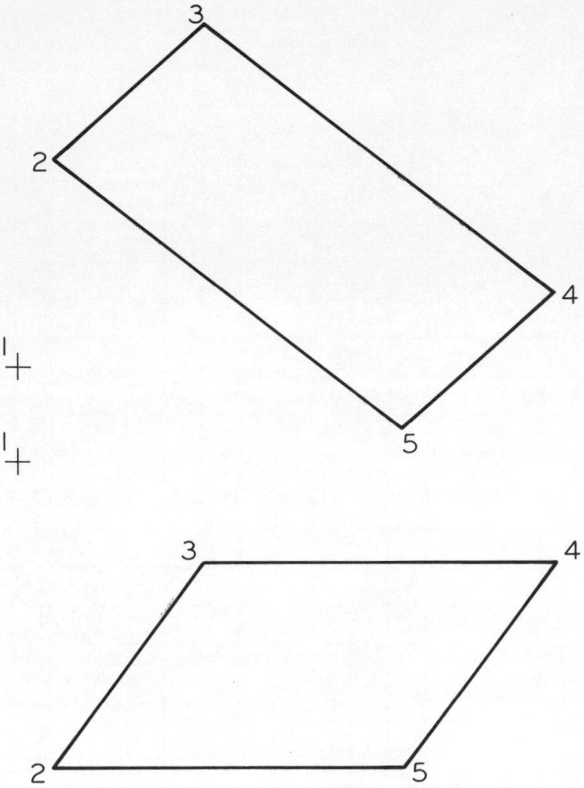
3
2
4
1
5
1
3
4
2
5

(2) Measure the length and show the views of the altitude of all cones having point V as the vertex and with their bases in plane 1-2-3.

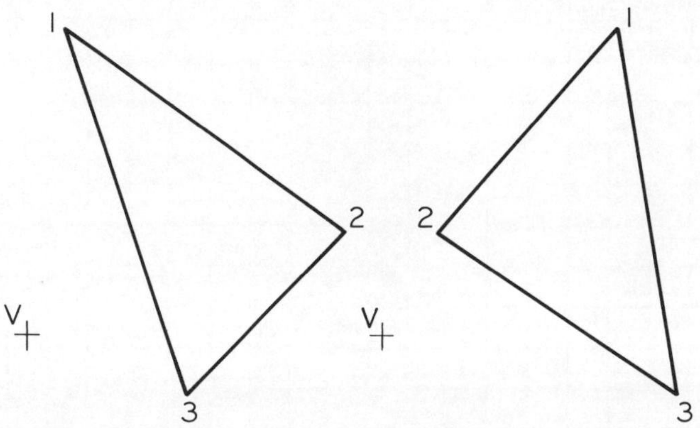
1
2
V
3
1
2
V
3

(3) The axis of a right pyramid lies along line 1-2. The vertex is at point V. The base is an equilateral triangle with one corner at point 3. Complete all views.

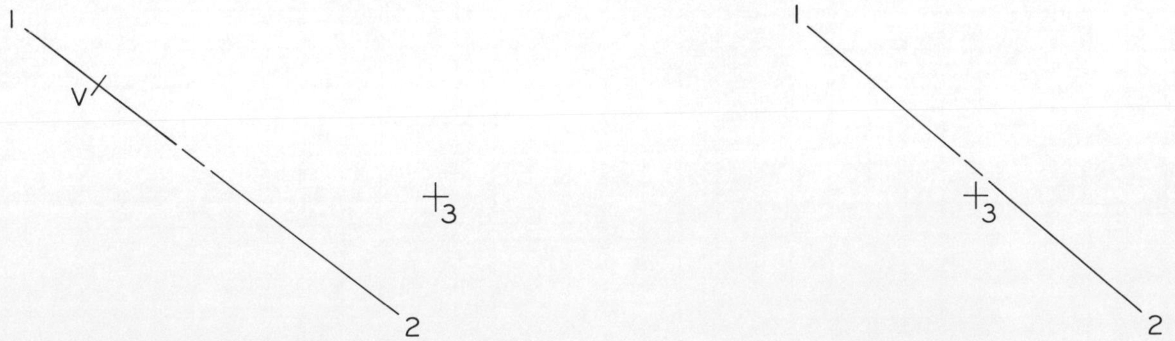
1
V
3
2
1
3
2

LINES AND PLANES PERPENDICULARITY	DRAWN BY	FILE NO.	DRAWING 65

	DRAWN BY	FILE NO.	DRAWI

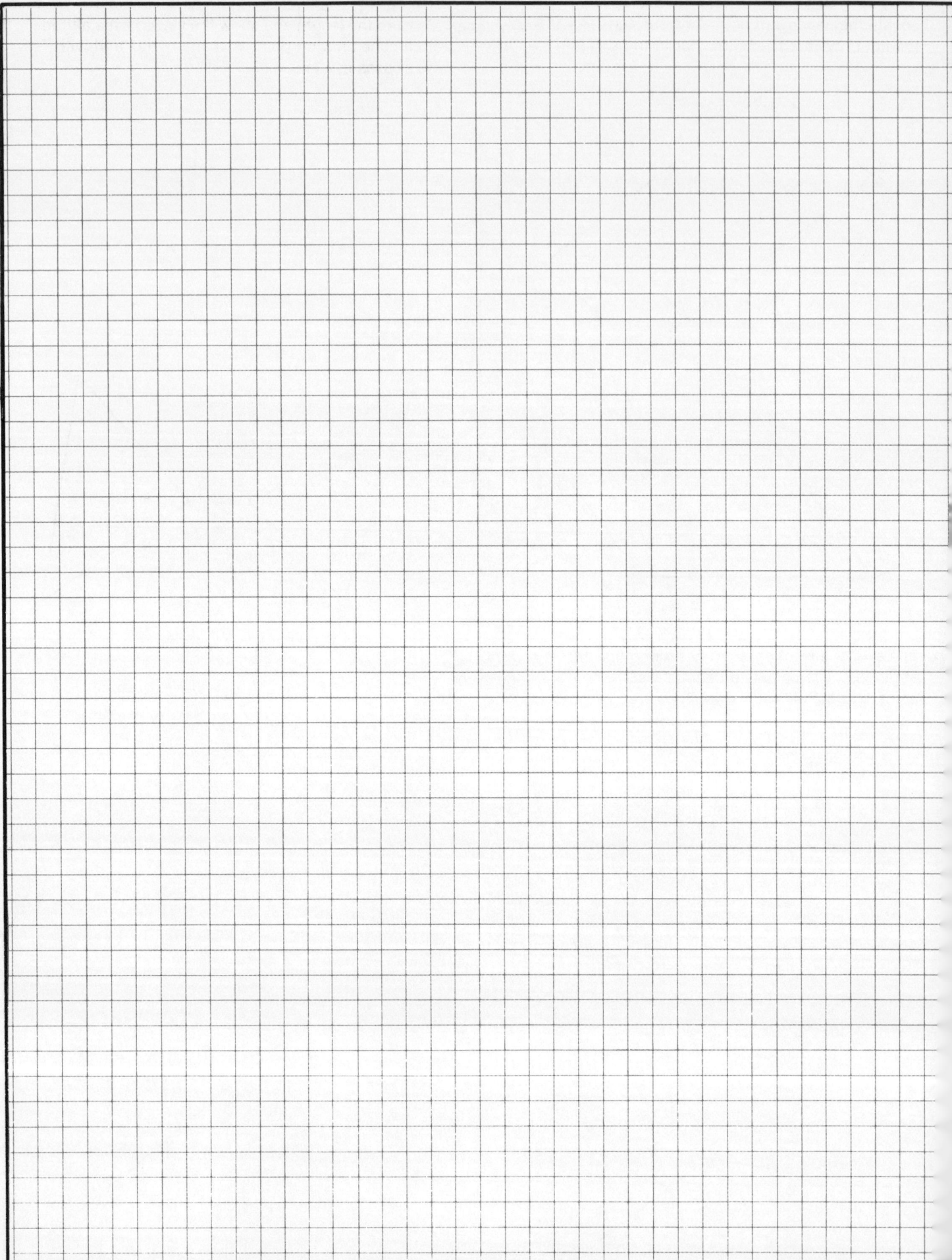

① Determine the bearing and length of the shortest ventilating shaft connecting tunnel 3-4 to tunnel 1-2. Scale: 1/5000.

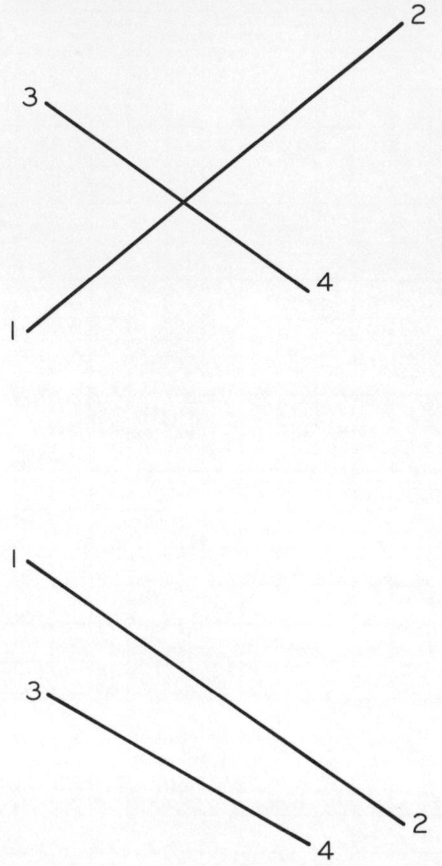
2
3
4
1
1
3
2
4

② High-frequency lead 1-2 must clear lead 3-4 and corner 5 of an electronic component by a minimum of 10 mm. If constructed as per the given layout, will the unit pass inspection?

YES________

NO ________

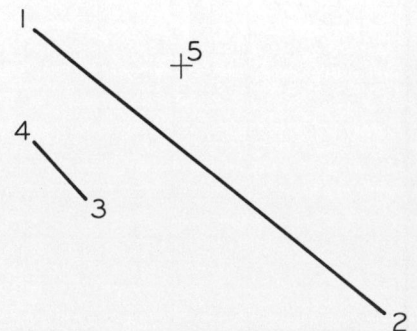

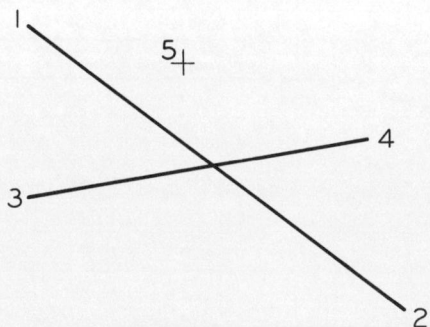

1
5
4
3
2
1
5
4
3
2

	DRAWN BY	FILE NO.	DRAWI

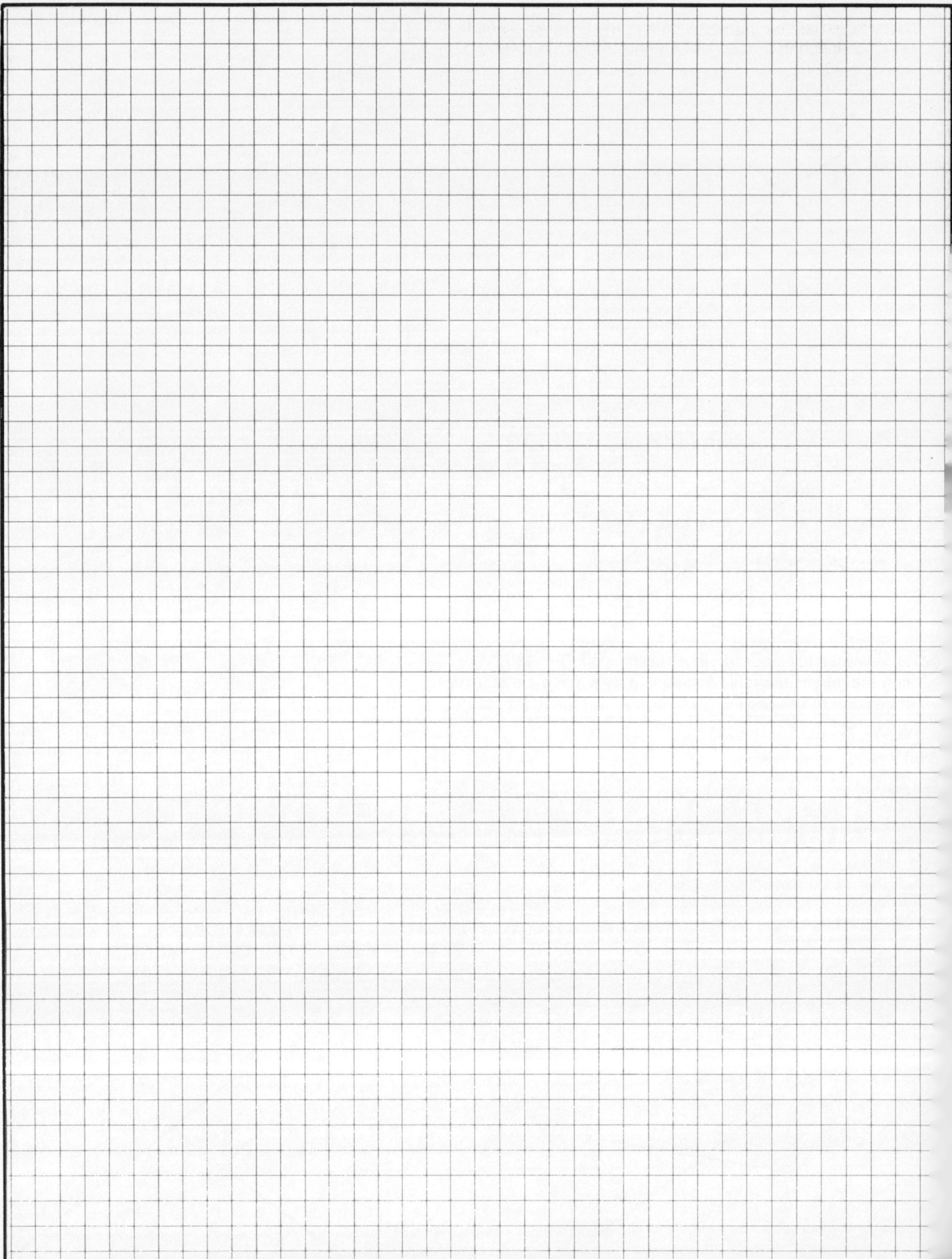

(1) Connect pipes 1-2 and 3-4 with the shortest branch parallel to the side (profile) wall.
Determine the length and show the views of the branch.
Scale: 1/100.

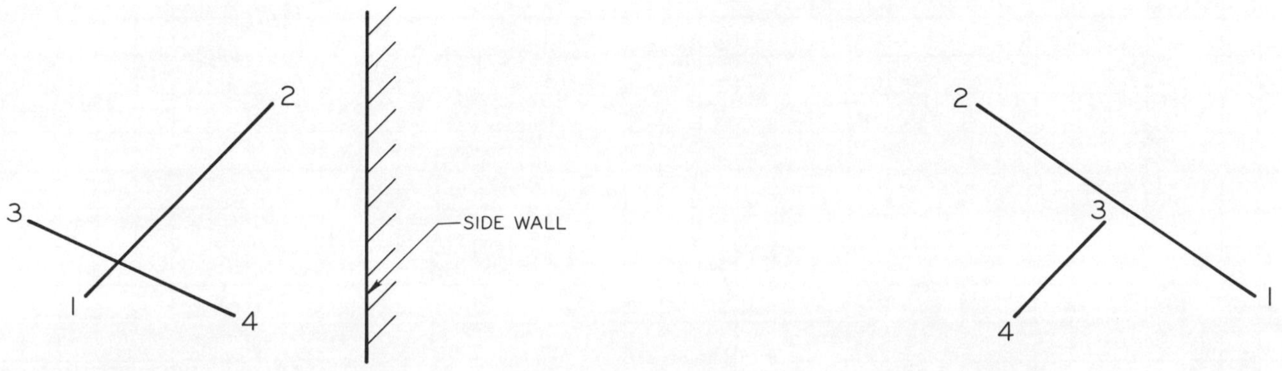

2
3
1
4
SIDE WALL
2
3
1
4

(2) Ski slope 1-2 is connected to ski slope 3-4 with the shortest path having a grade of − 10%.
Find the length and bearing and show the views of the path.
Scale: 1/5000.

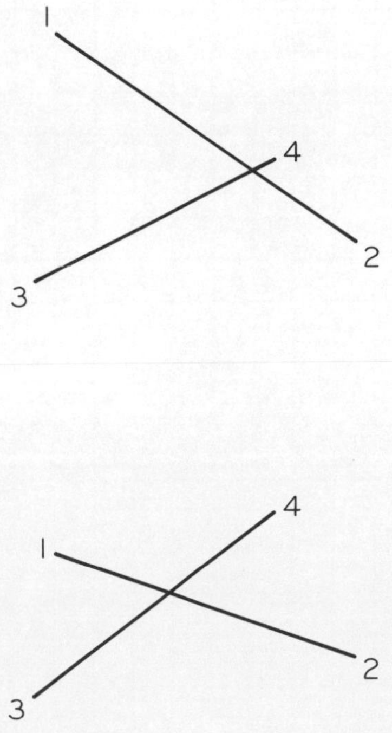

1
4
2
3
4
1
2
3

LINES AT SPECIFIED ANGLES	DRAWN BY	FILE NO.	DRAWING
SKEW LINES			67

	DRAWN BY	FILE NO.	DRAWI

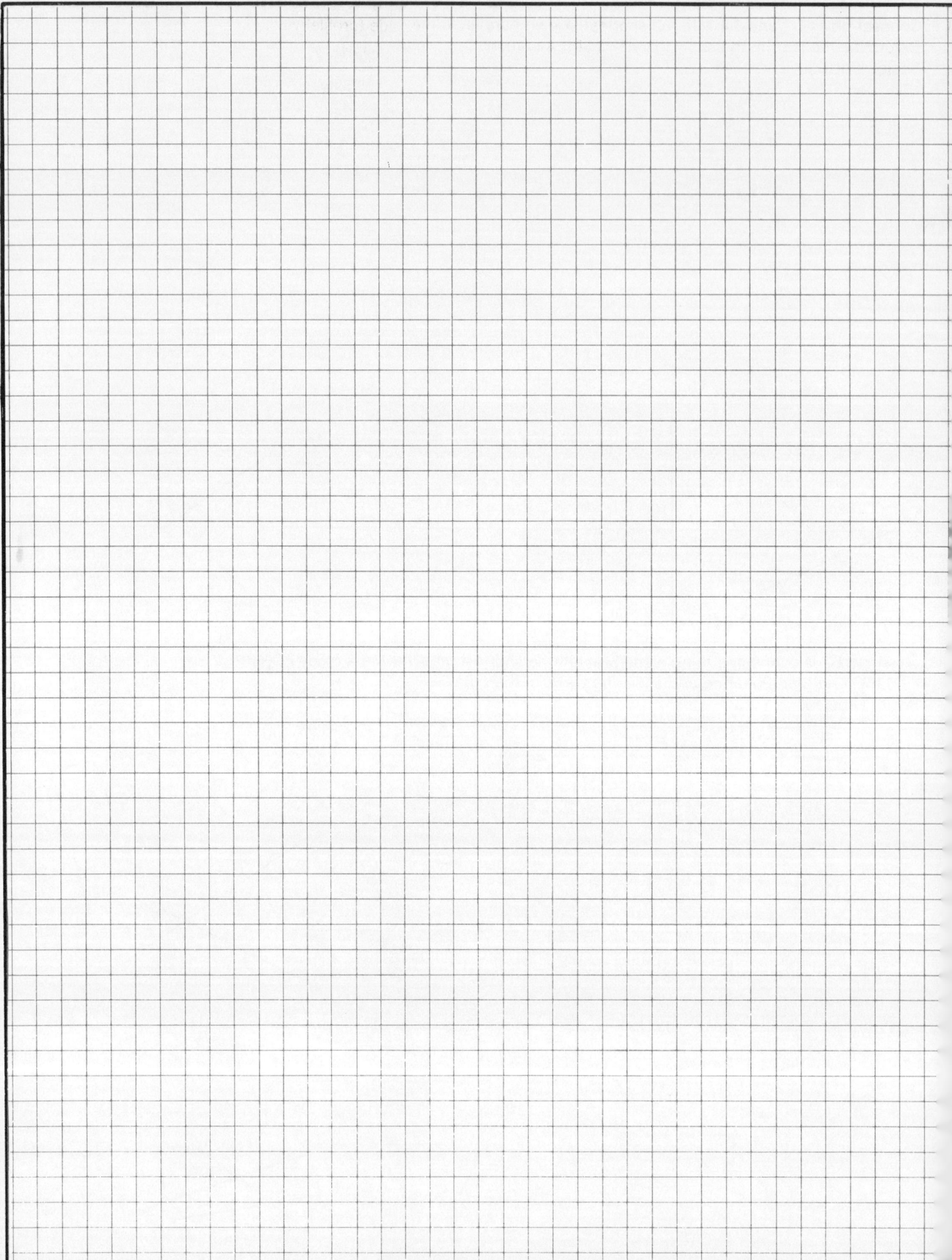

(1) Establish the intersection of unlimited plane 1-2-3 with the pyramid.

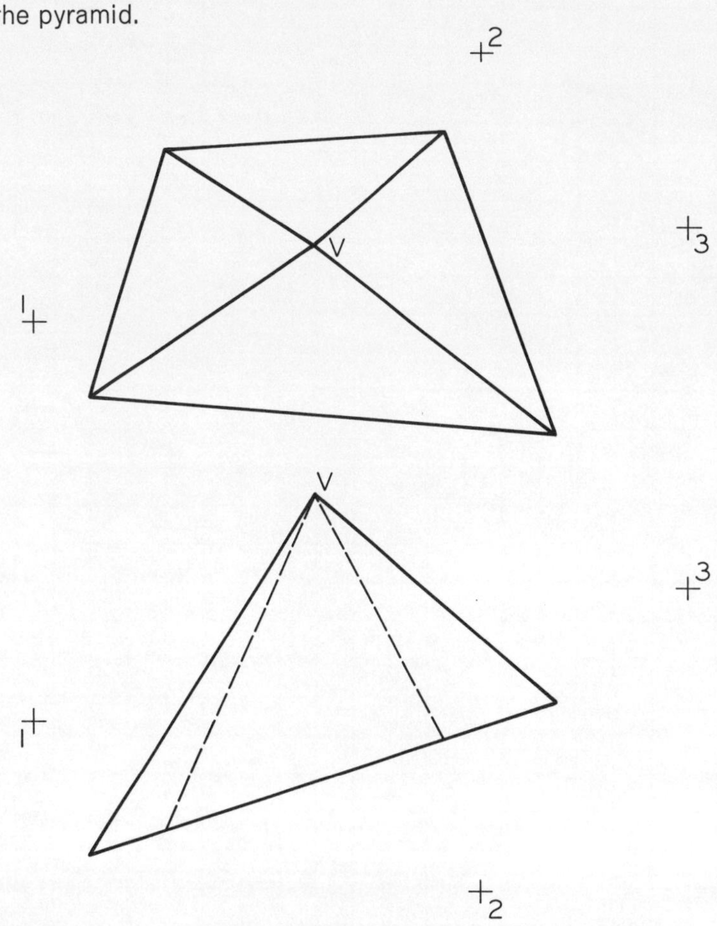

2
V
3
1
V
3
1
2

(2) Determine the intersection of limited plane 1-2-3-4 and the prism.

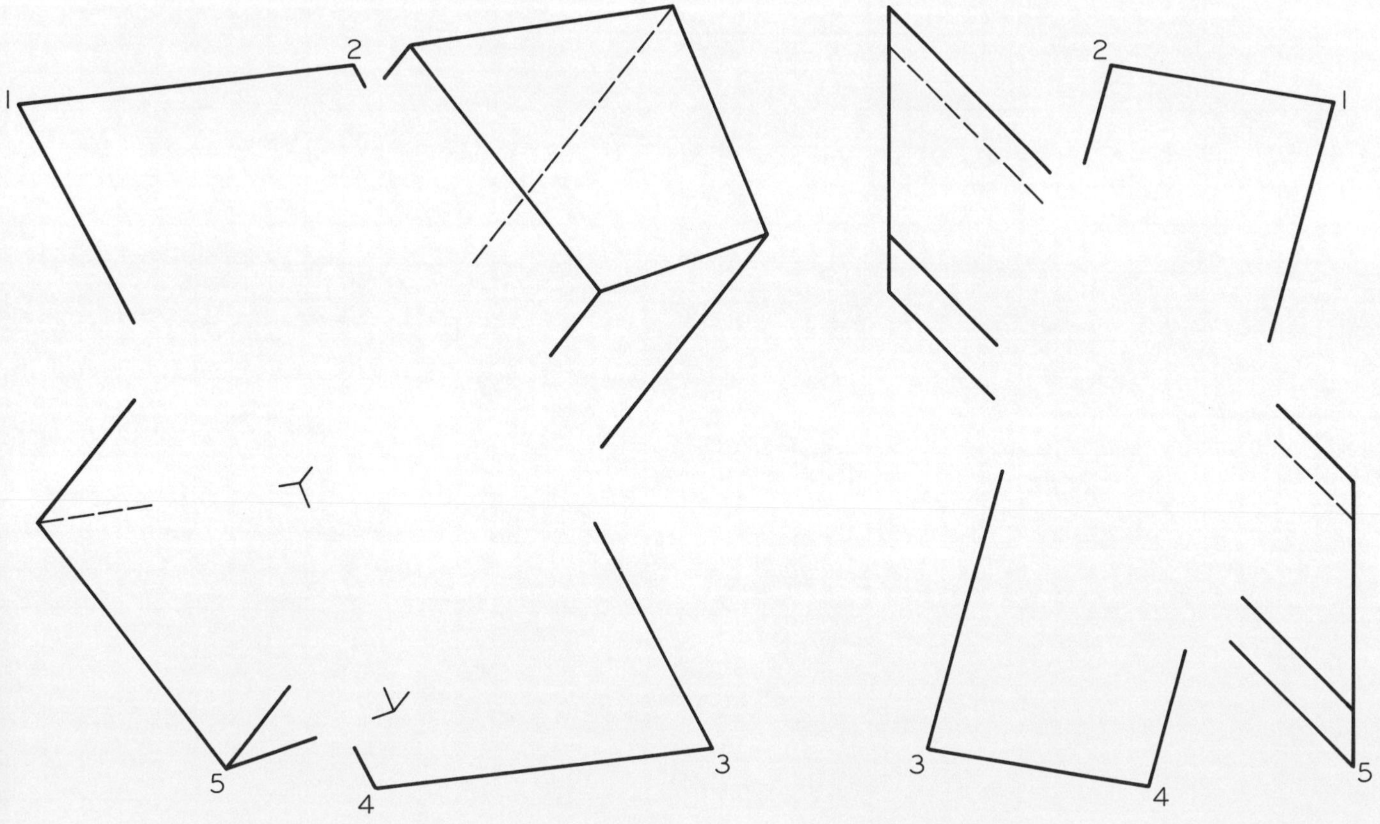

2
1
5
4
3
2
1
3
4
5

PLANES AND POLYHEDRA INTERSECTIONS	DRAWN BY	FILE NO.	DRAWING 68

	DRAWN BY	FILE NO.	DRAWI

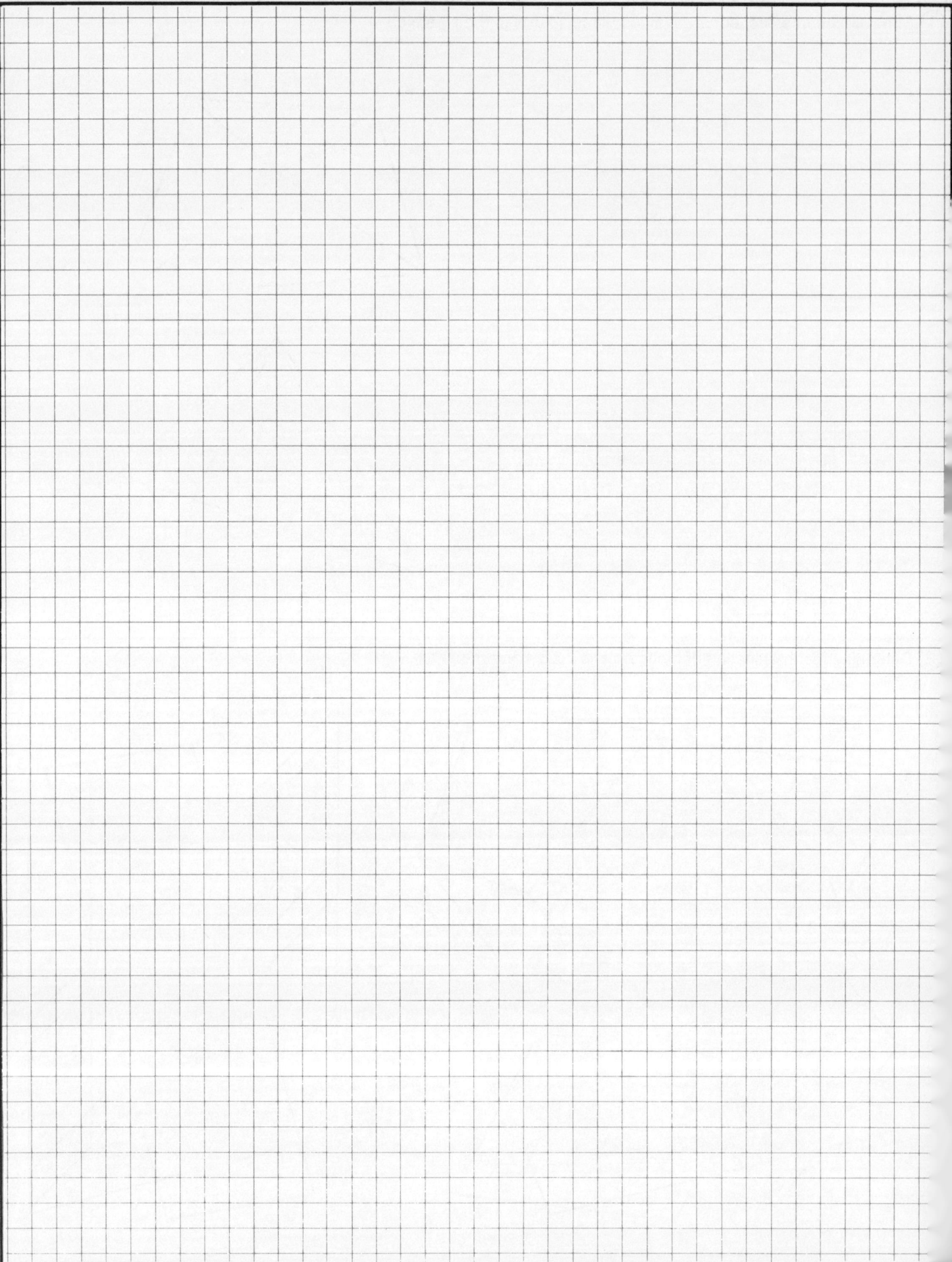

(1) Show the intersection of the cylinder and unlimited plane 1-2-3.

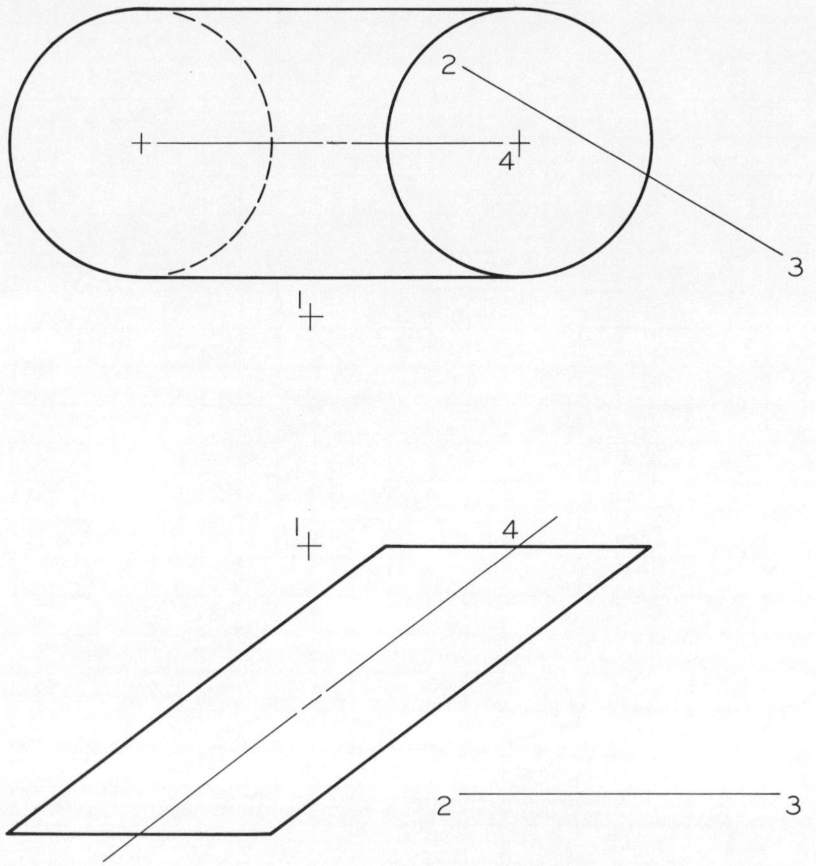
2
4
3
1
1
4
2
3

(2) Establish the intersection of the frustum of the cone and the unlimited plane 1-2-3.

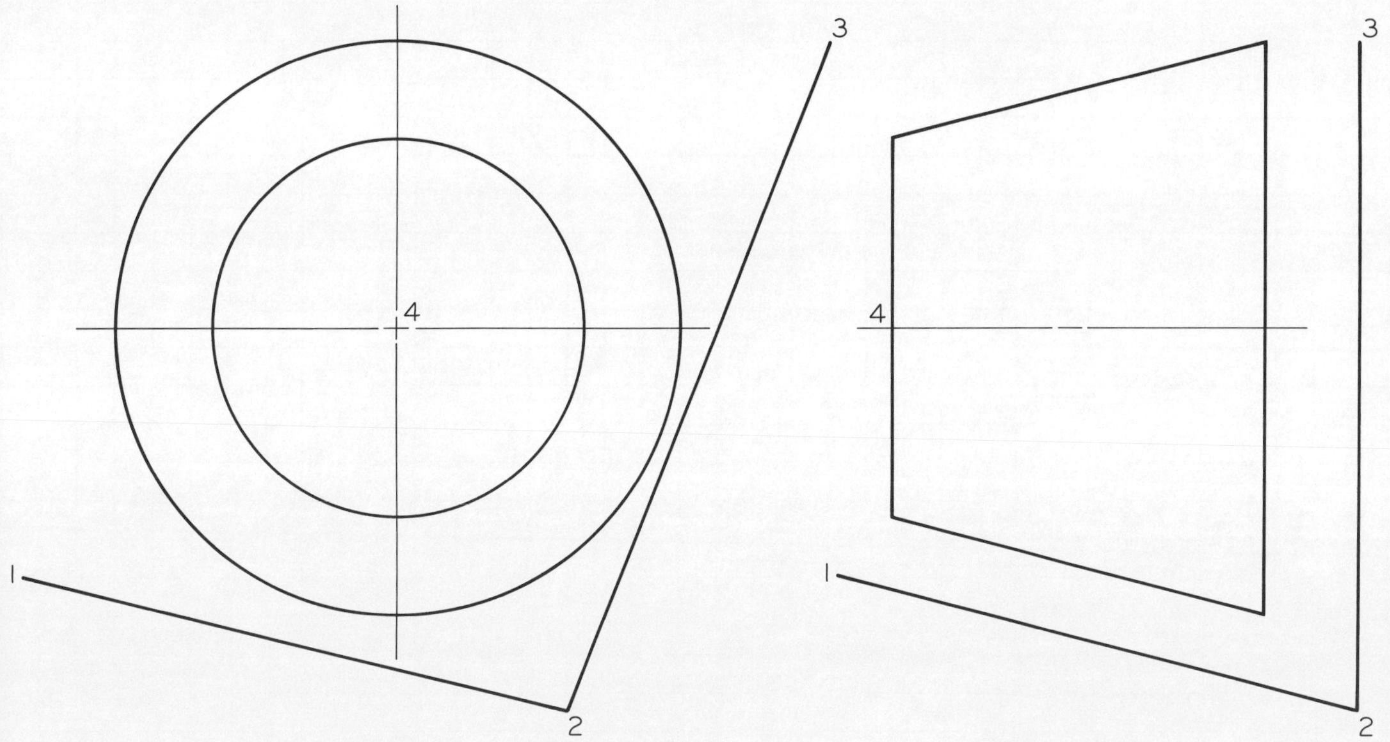
3
3
4
4
1
1
2
2

	DRAWN BY	FILE NO.	DRAWI

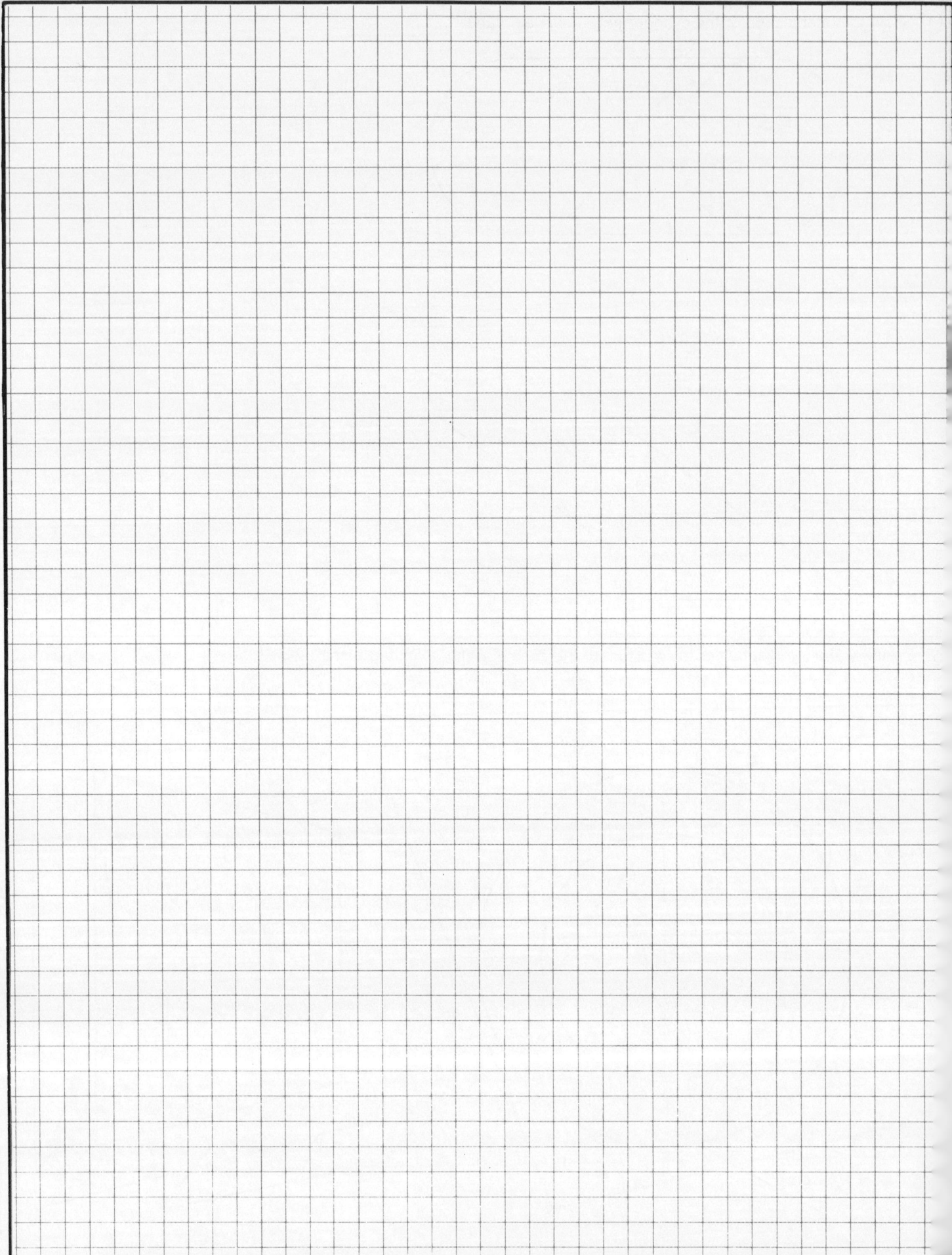

(1) Establish the figure of intersection of the two prisms.

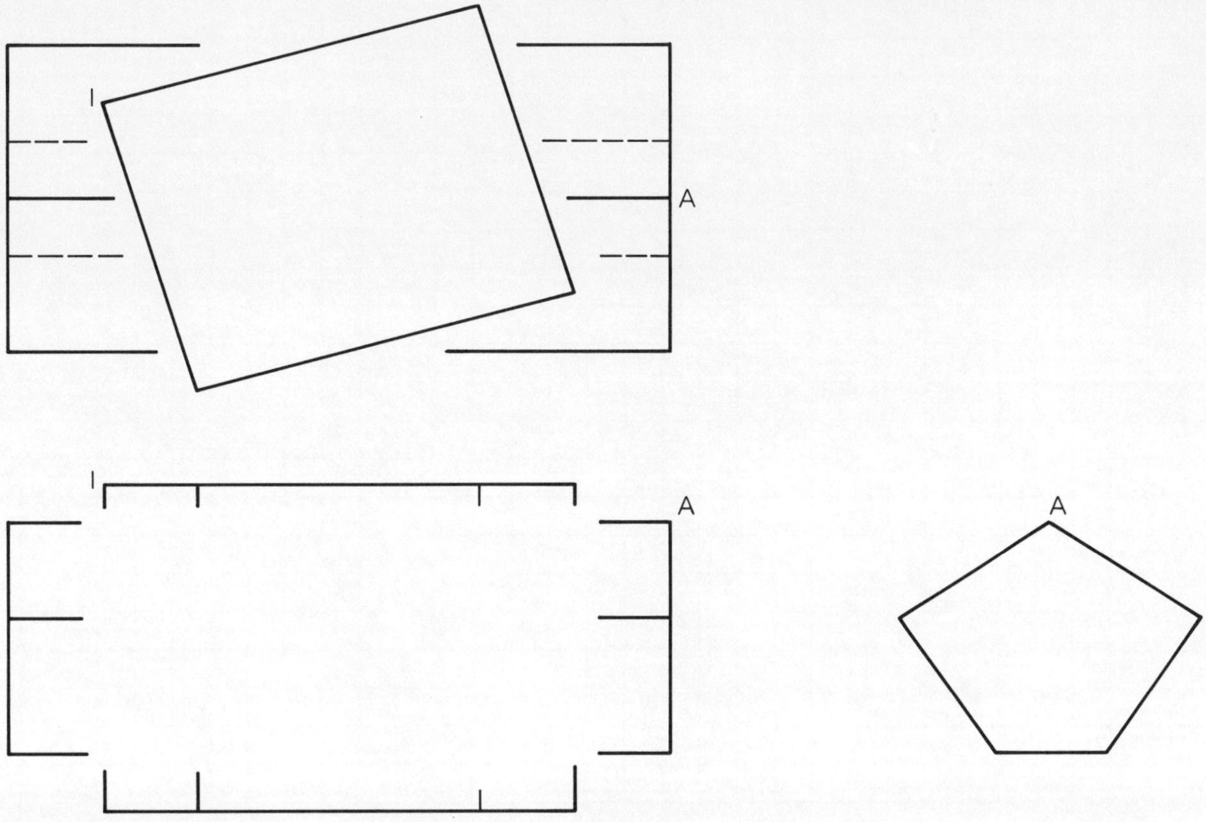
I
A
I
A
A

(2) Find the figure of intersection of the prism and pyramid.

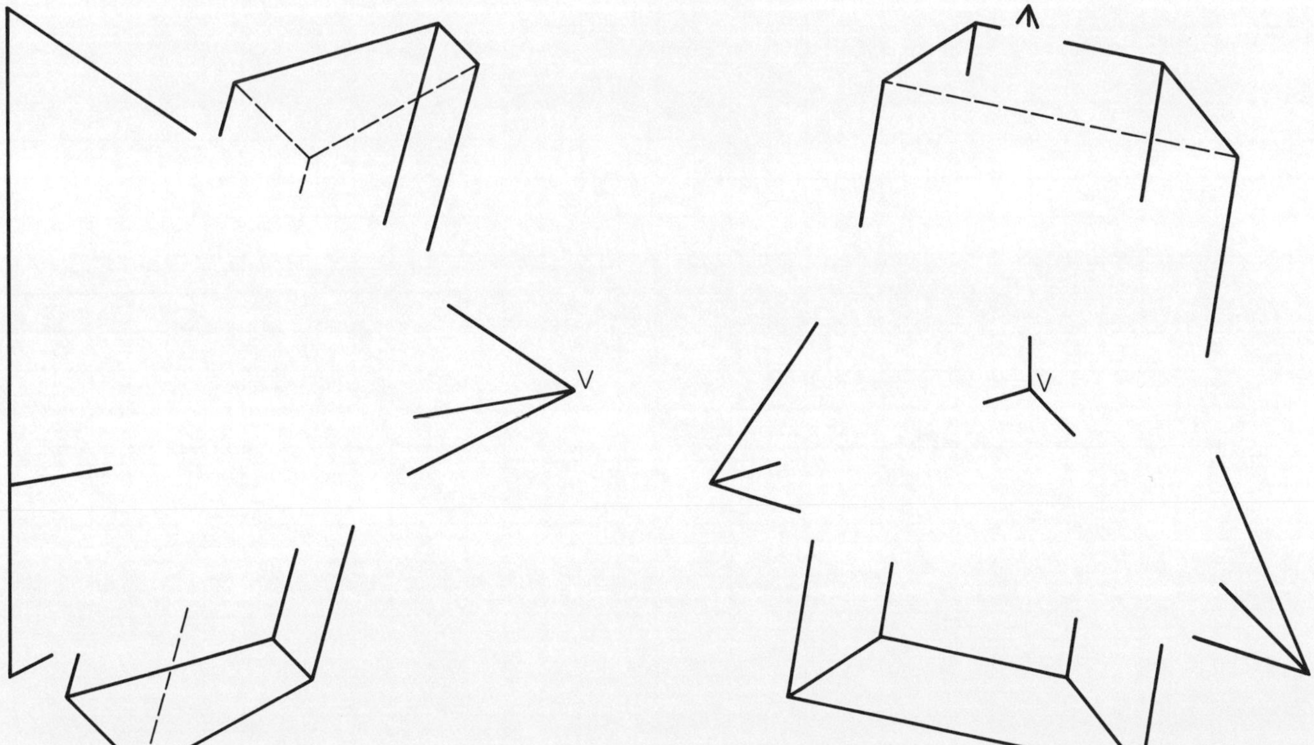
V
V
I
I

PRISMS AND PYRAMIDS INTERSECTIONS	DRAWN BY	FILE NO.	DRAWING 70

	DRAWN BY	FILE NO.	DRAWI

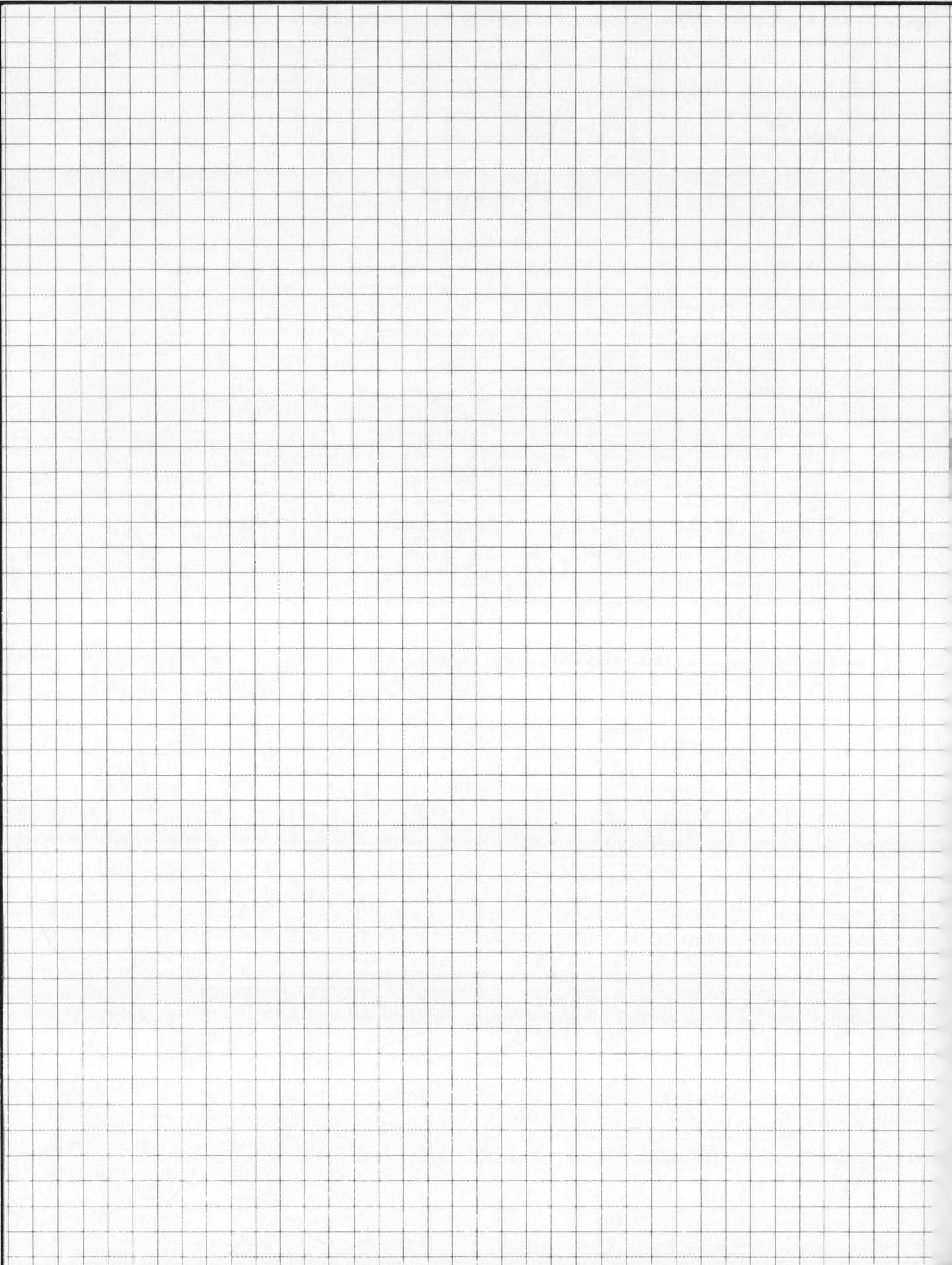

Complete the views of the cone and cylinder, including the figure of intersection.

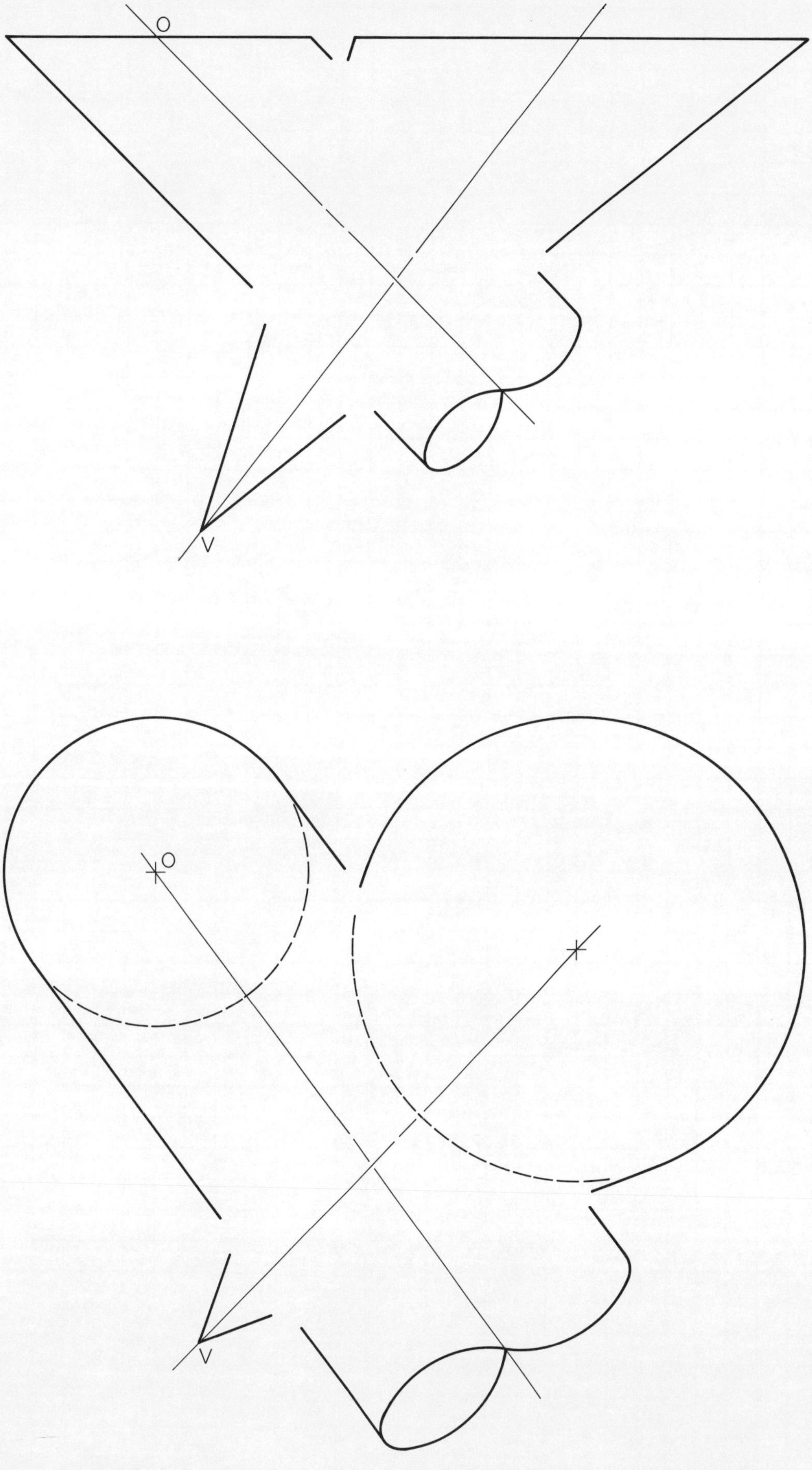
O
V
O
V

OBLIQUE CONE AND CYLINDER INTERSECTIONS	DRAWN BY	FILE NO.	DRAWING 72

	DRAWN BY	FILE NO.	DRAW

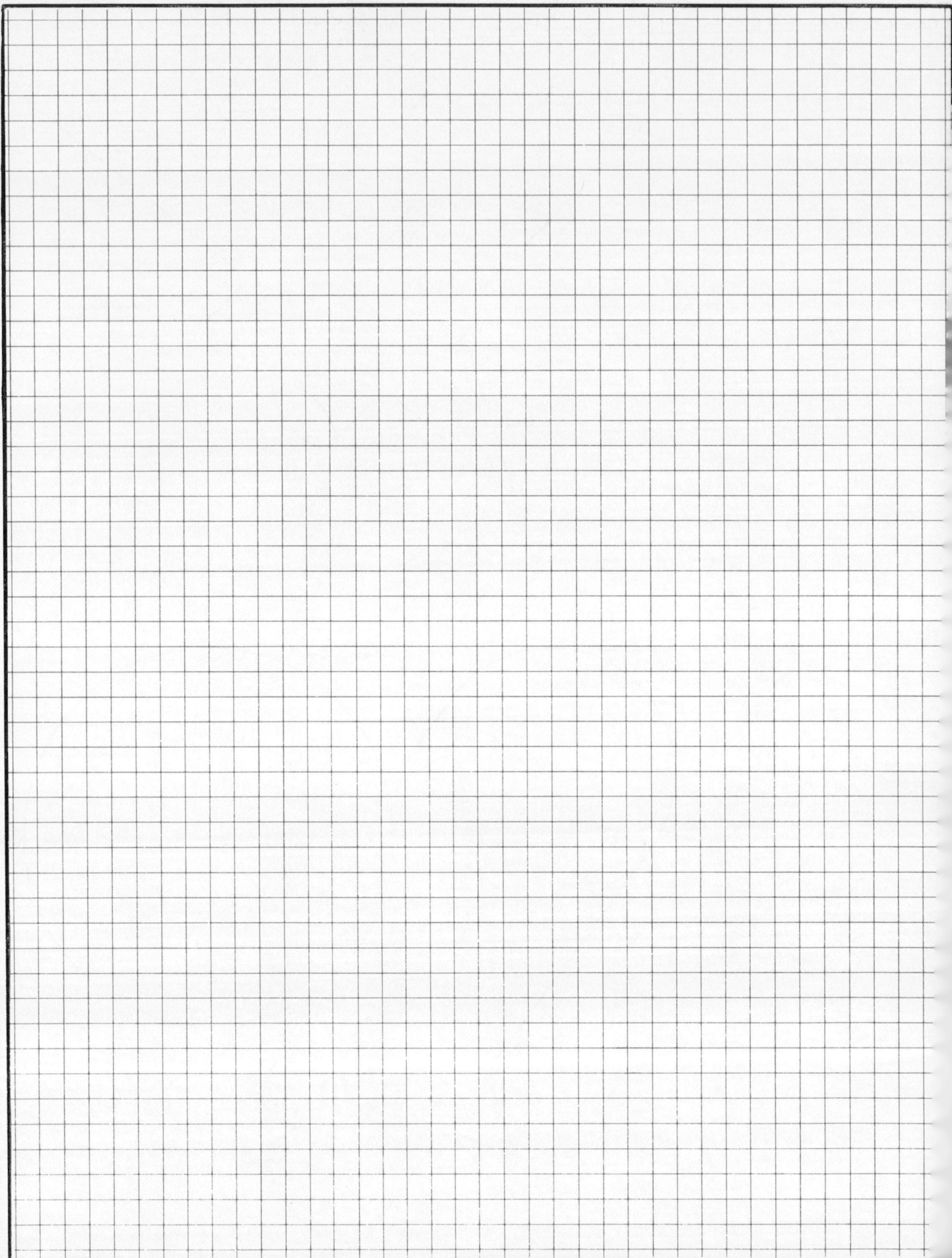

(1) Develop the lateral surface of the sheet metal prism, starting with edge EF.

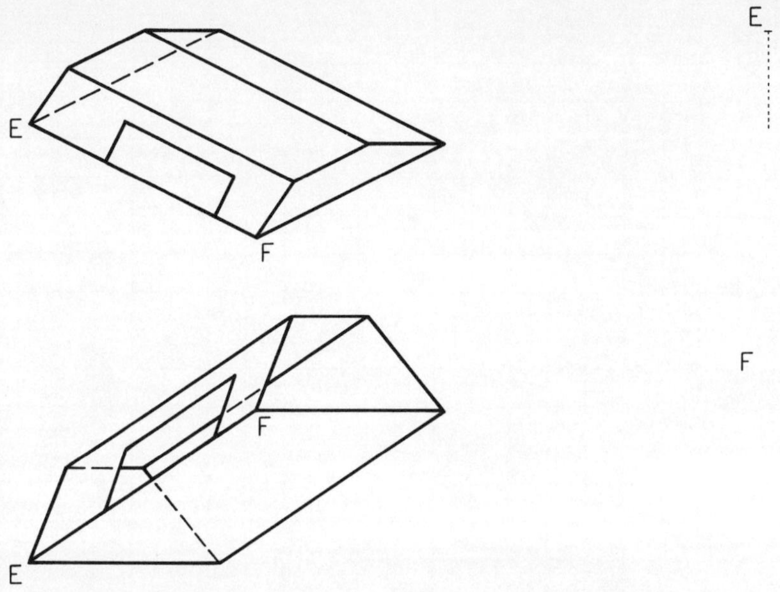

E
F
E
F
E
F

(2) Develop the lateral surface of the cylindrical duct, starting the development as indicated.

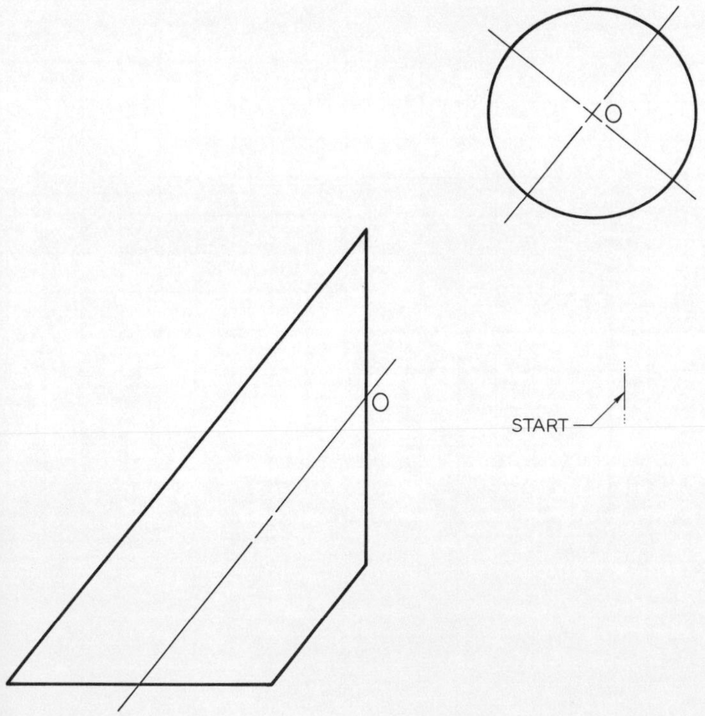

O
O
START

PRISM AND CYLINDER ARALLEL-LINE DEVELOPMENT	DRAWN BY	FILE NO.	DRAWING 73

	DRAWN BY	FILE NO.	DRAWI

① Develop the lateral surfaces of the pyramidal transition piece.

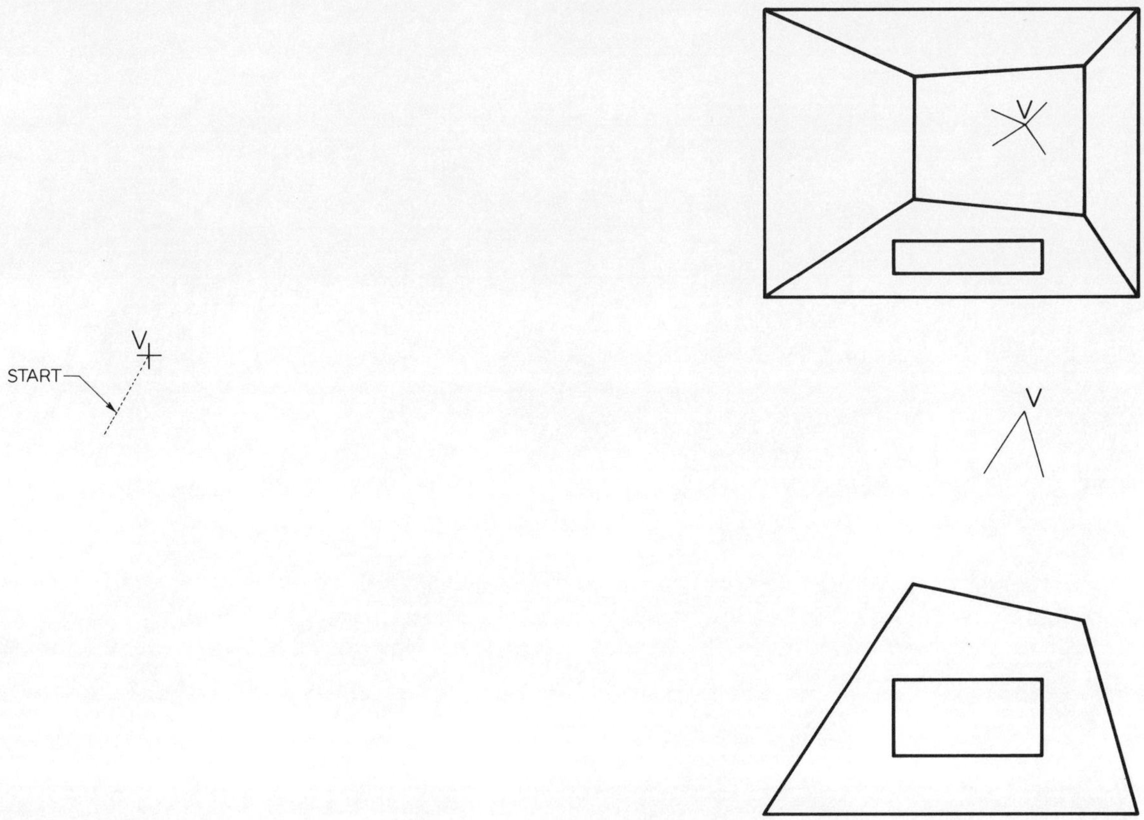
V
START
V
V

② Lay out a half development of the truncated right-circular cone.

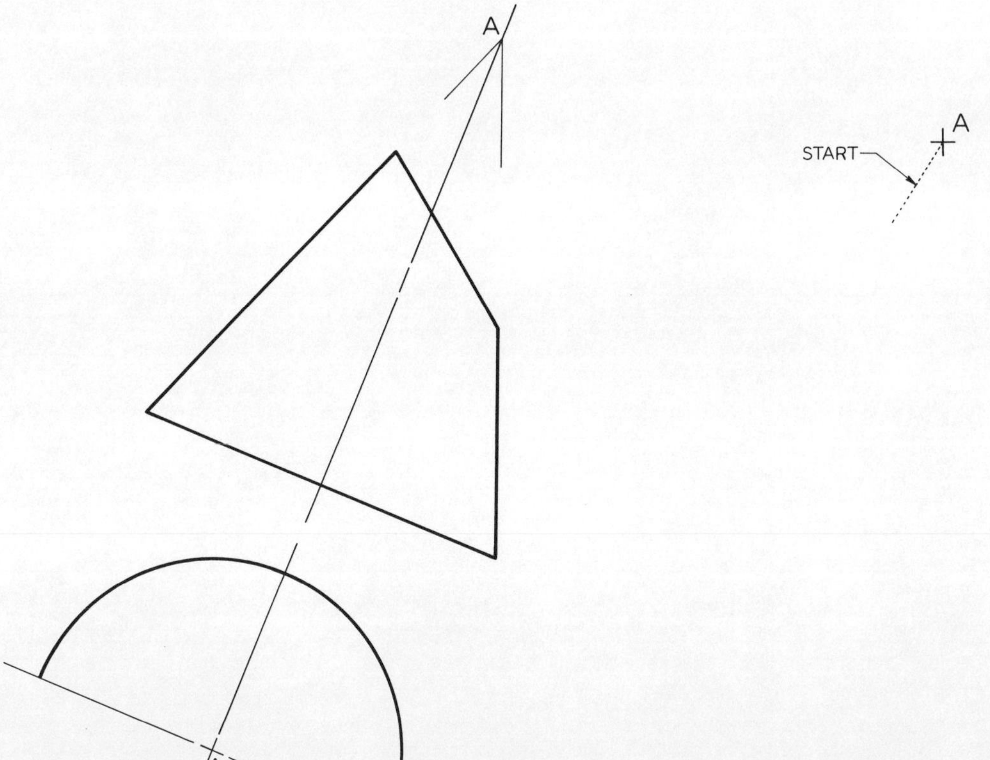
A
START
A
A

PYRAMID AND CONE RADIAL-LINE DEVELOPMENT	DRAWN BY	FILE NO.	DRAWING 74

	DRAWN BY	FILE NO.	DRAWI

(1) Pass a plane tangent to the cone and containing point I on the surface of the cone.

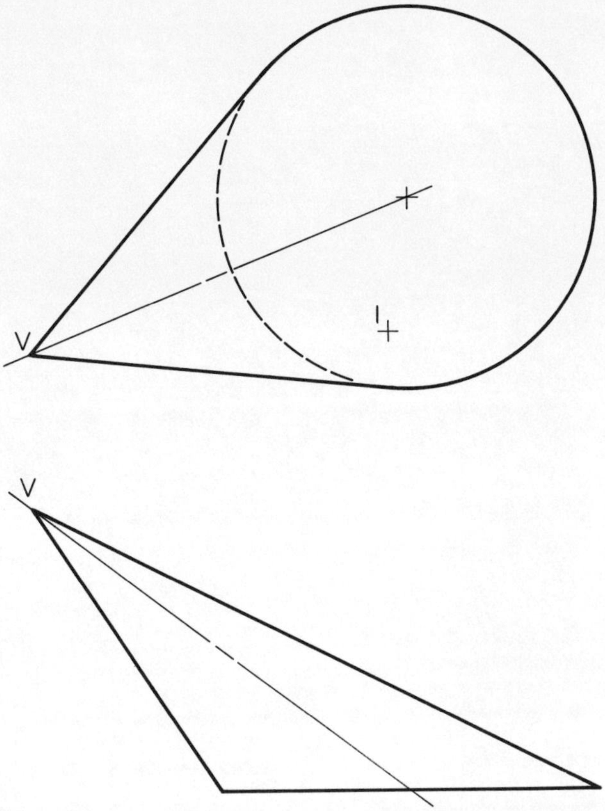

V
I
V

(2) Establish a plane tangent to the cylinder and containing point I.

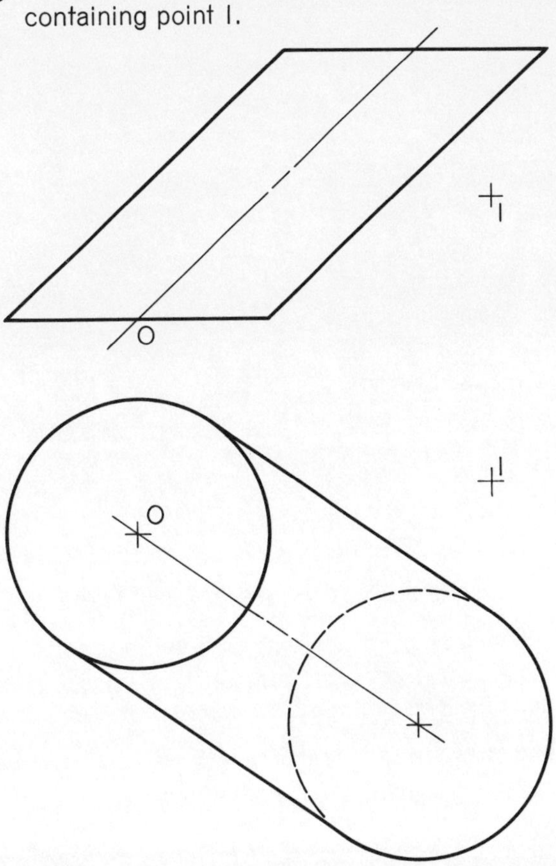

I
O
I
O

(3) Show a plane tangent to the cylinder and parallel to line I-2.

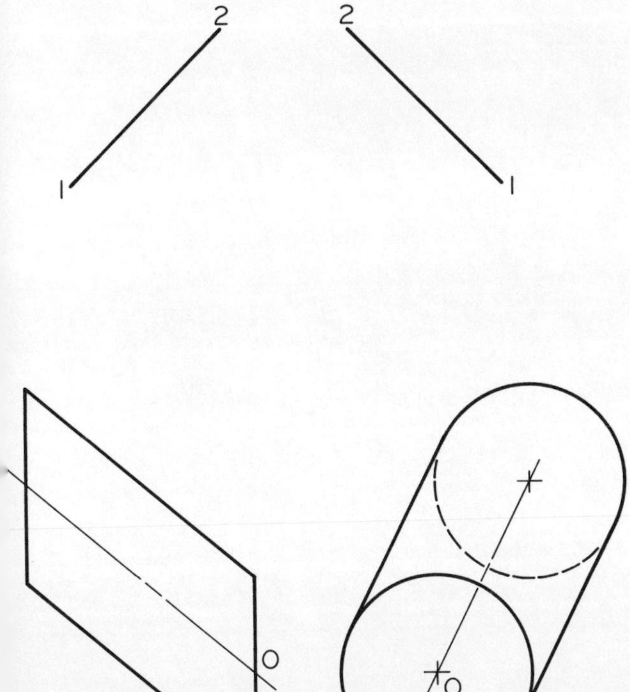

2
I
2
I
O
O

(4) Draw three lines tangent to the sphere at point I on its surface.

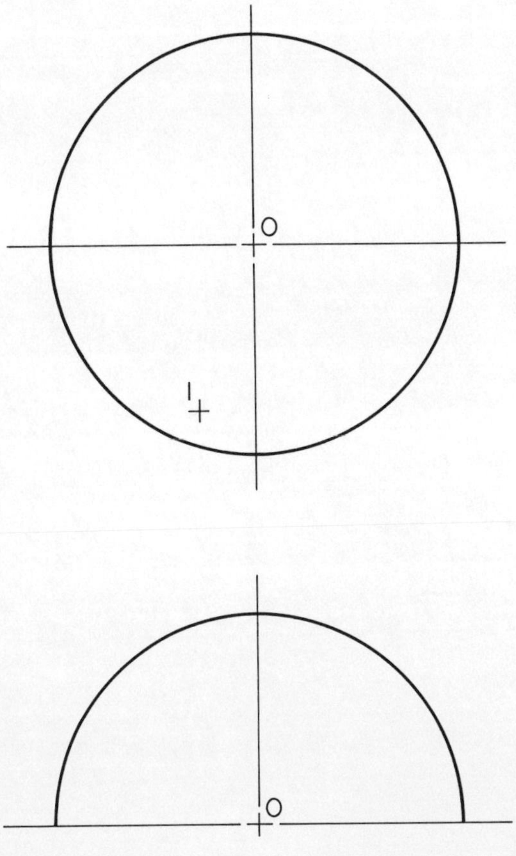

O
I
O

LINE AND PLANE TANGENCIES	DRAWN BY	FILE NO.	DRAWING 76

	DRAWN BY	FILE NO.	DRAWIN

(1) Represent a plane containing line 1-2 and making an angle of 60° with a frontal plane.

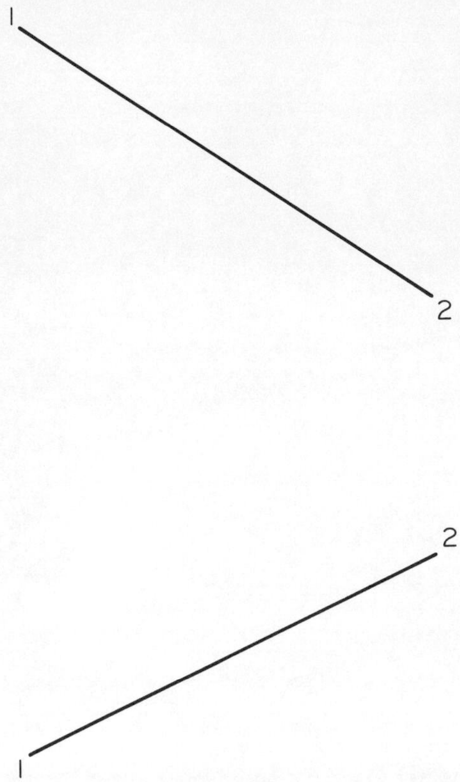

1
2
2
1

(2) Pass a plane through line 3-4 and making an angle of 110° with the inclined surface.

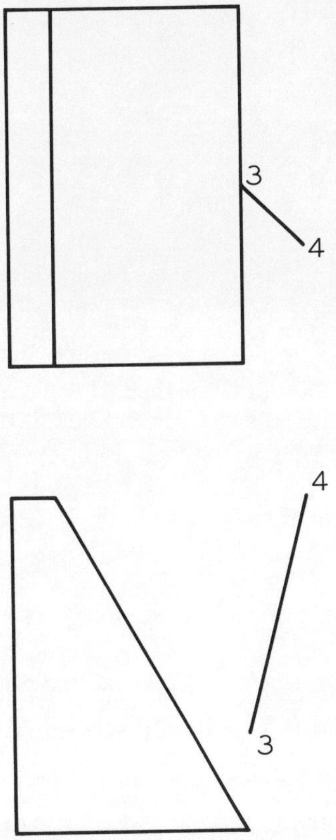

3
4
4
3

(3) Complete the views of line 5-6 which forms angles of 35° with a frontal plane and 45° with a horizontal plane.

5
5

(4) Point 4 is in plane 1-2-3. Find a 25mm line 4-5 which forms an angle of 30° with a frontal plane and lies within triangle 1-2-3.

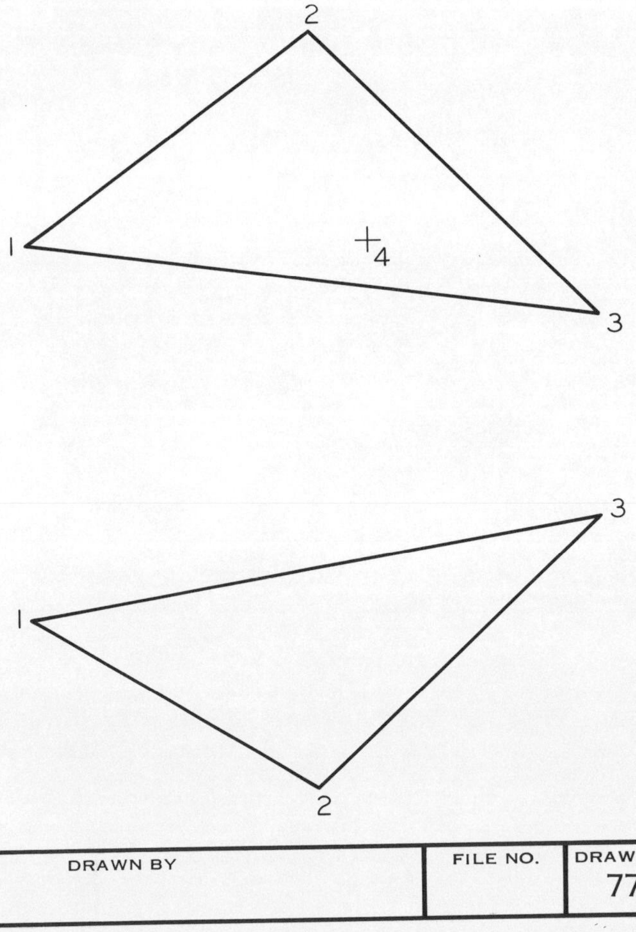

2
1
4
3
3
1
2

SPECIFIED ANGLES TANGENCIES	DRAWN BY	FILE NO.	DRAWING 77

	DRAWN BY	FILE NO.	DRAWING

	A	B	C	D	E	F	G	H
7	116.0	104.0	107.2	116.2	131.1	136.3	143.7	156.3
6	118.5	102.2	114.7	127.3	132.1	148.7	153.6	151.6
5	122.5	120.0	117.4	128.6	141.2	155.8	166.3	145.1
4	133.3	131.0	128.5	131.7	148.0	157.2	155.3	143.6
3	135.3	148.4	147.2	152.0	148.3	146.5	142.8	141.9
2	146.1	156.7	138.2	143.2	155.6	134.7	132.8	127.4
1	155.3	137.2	142.1	156.8	145.3	136.2	126.5	122.5

TOPOGRAPHIC MAP AND PROFILES TOPOGRAPHY	DRAWN BY	FILE NO.	DRAWING 78

	DRAWN BY	FILE NO.	DRA

(1) Find the strike and dip of upper bedding plane 1-2-3 of a stratum.

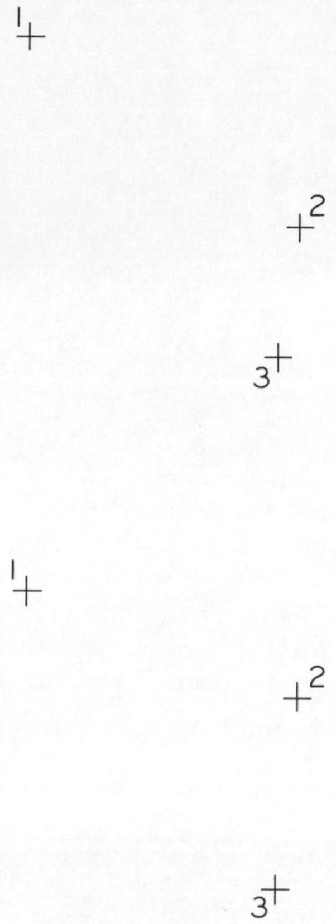

1
2
3
1
2
3

(2) Line 1-2 lies in a bedding plane of a seam having a strike of N65°E and a dip of 35°NW. Complete the top view.

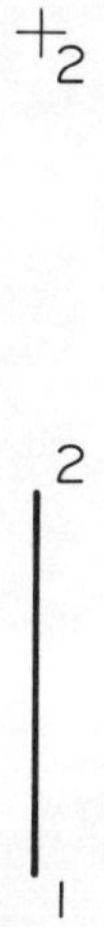

2
2
1

(3) Plane 1-2-3 is the upper bedding plane of a vein. Point 4 lies in the parallel lower bedding plane. Find the strike, dip, and thickness. Scale: 1/500.

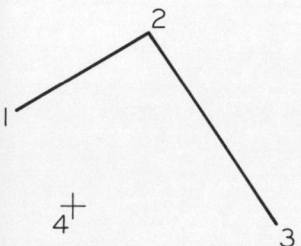

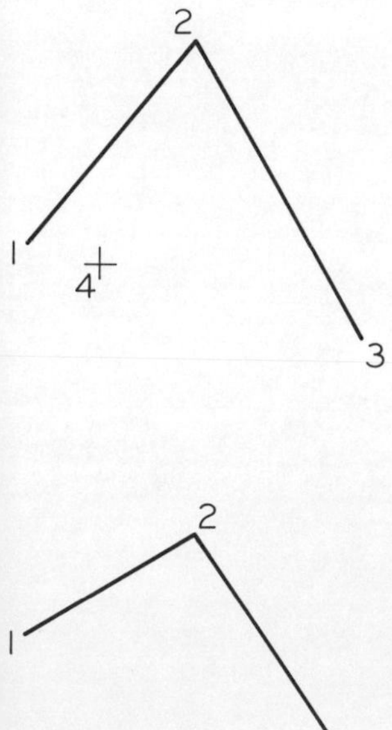

2
1
4
3
2
1
4
3

(4) Test drillings located points 1 and 2 in a plane having a dip of 55°NE. Find the strike of the plane.

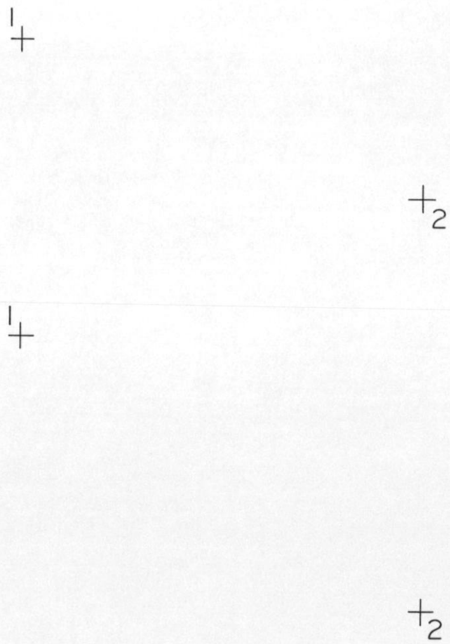

1
2
1
2

STRIKE, DIP, AND THICKNESS MINING AND GEOLOGY	DRAWN BY	FILE NO.	DRAWING 79

	DRAWN BY	FILE NO.	DRAW

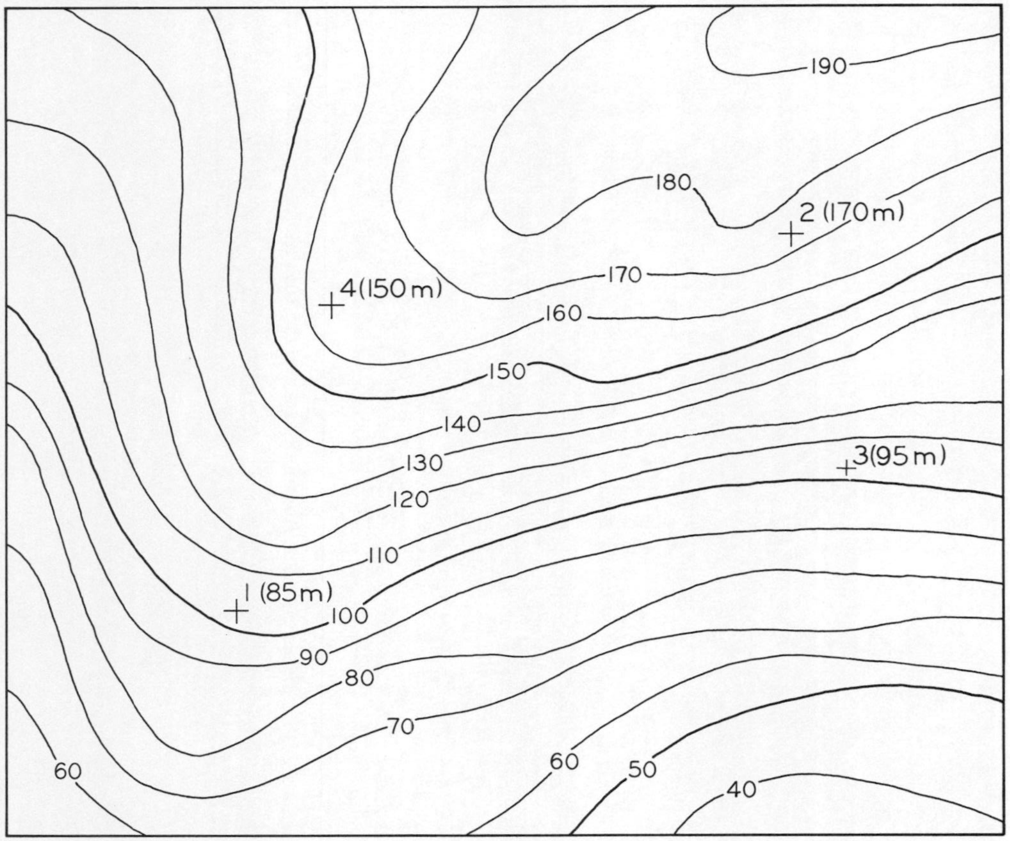

190
180
2 (170m)
170
4(150m)
160
150
140
3(95m)
130
120
110
1 (85m)
100
90
80
70
60
50
40
60

180
160
140
120
100
80

Test drillings located points 1, 2, and 3 in the upper bedding plane of a stratum. Point 4 is in the parallel lower bedding plane.

a. Find the strike, dip, and thickness of the stratum. Scale: 1/2000.
b. Plot the outcrop lines of the bedding planes.

	DRAWN BY	FILE NO.	DRAWI

Establish the views of the vertices of spherical triangle ABC, given that angle A = 110°, side b = 40°, and side c = 60°. Solve for angles B and C, and side a.

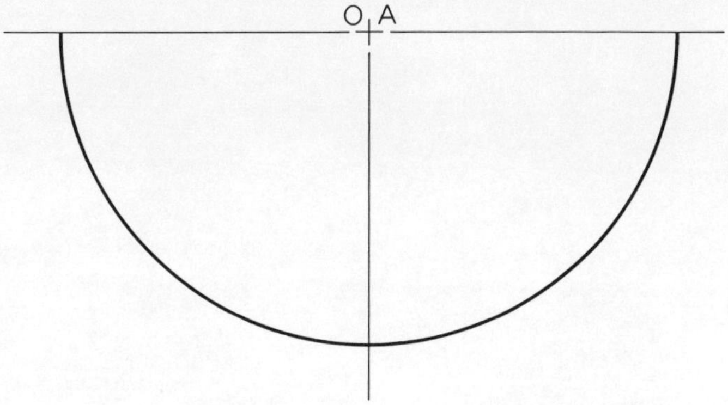

O
A

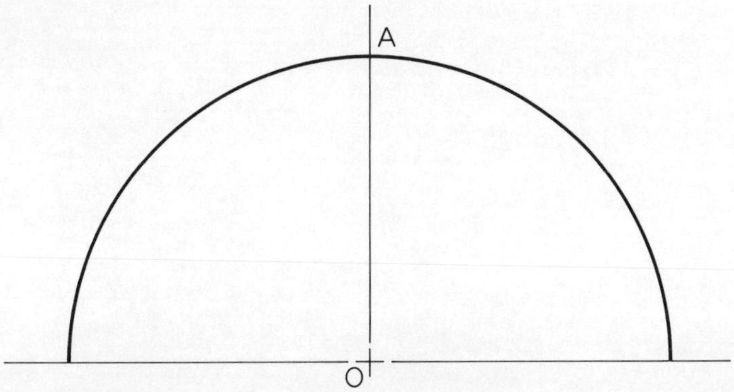

A
O

	DRAWN BY	FILE NO.	DRAW

nd, in nautical miles, the great-circle distance on the surface of the earth from location A, latitude 50°N, ngitude 20°E, to location B, latitude 15°S, longitude 130°W. Determine the initial and final azimuth bearings of this course.

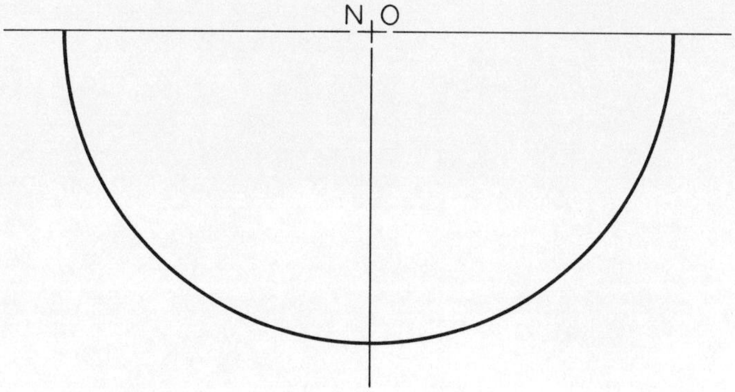

N
O

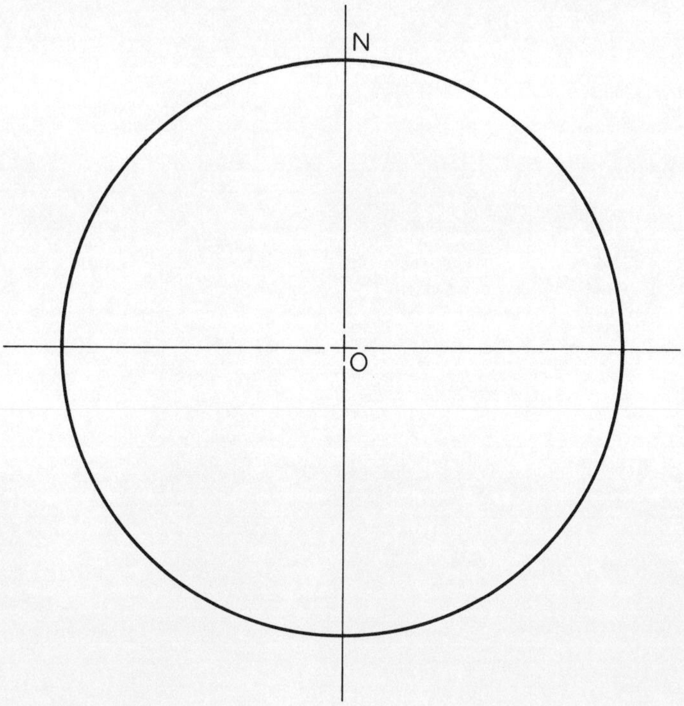

N
O

	DRAWN BY	FILE NO.	DRAW

(1) Find the vertical and horizontal components of the resultant of the force system shown at P. Start the force polygon at p. Force scale: 1mm = 1N.

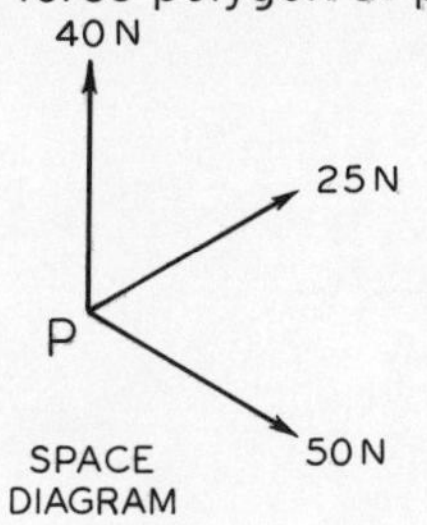
40 N
25 N
P
50 N
SPACE DIAGRAM

p

VECTOR DIAGRAM

(2) Starting the force diagram at a, find the stress in the crane members. Force scale: 1mm = 10N. Find the reactions R_1 and R_2.

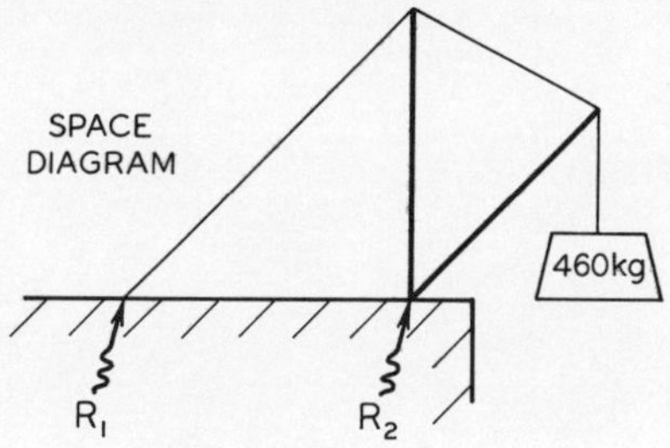
SPACE DIAGRAM
460kg
R_1
R_2

a

VECTOR DIAGRAM

(3) An aircraft at A is in level flight on a compass course of N30°; but because of a crosswind from due west of 110 knots it actually passes over point T. Determine the air speed and true ground speed of the craft. Vector Scale: 1mm = 4 knots.

T

A

(4) An aircraft carrier at A is cruising at 20 knots on a course of N180°. Another ship at B is traveling at 30 knots on a course of N270°. Find the positions of the ships and the distance between them at the moment of closest approach. Vector scale: 1″ = 20 knots. Distance scale: 1″ = 1000 yds.

A

B

CONCURRENT COPLANAR VECTORS VECTOR ANALYSIS	DRAWN BY	FILE NO.	DRAWING 84

	DRAWN BY	FILE NO.	DRAWI

Determine reactions R_1 and R_2, starting the force diagram at a_1. Find the loads in each member with a stress diagram starting at a_2. Scale: 1 mm = 100 N.

SPACE DIAGRAM

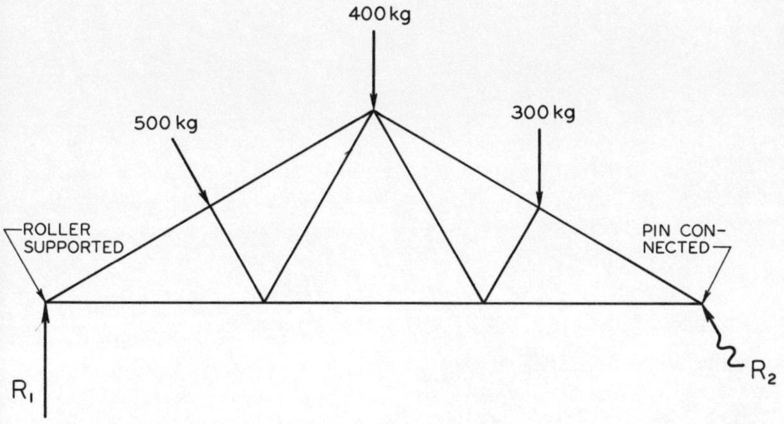

400 kg
500 kg
300 kg
ROLLER SUPPORTED
PIN CON-NECTED
R_1
R_2

FORCE DIAGRAM

a_1

STRESS DIAGRAM

a_2

NONCONCURRENT COPLANAR VECTORS VECTOR ANALYSIS	DRAWN BY	FILE NO.	DRAWING 85

	DRAWN BY	FILE NO.	DRAWI

(1) Find the stress in each tripod leg. Scale: 1/2000.

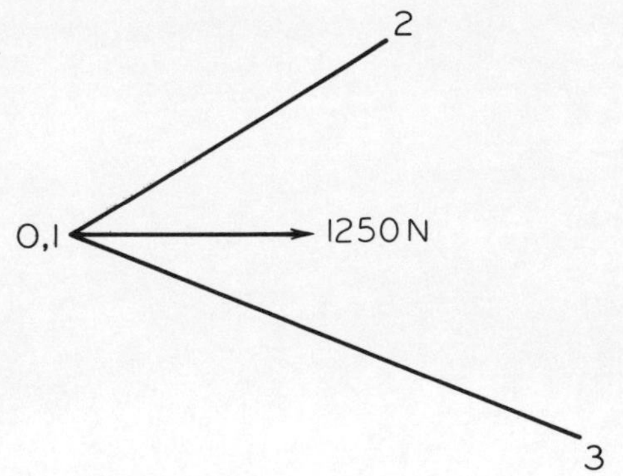

2
O,1
1250 N
3

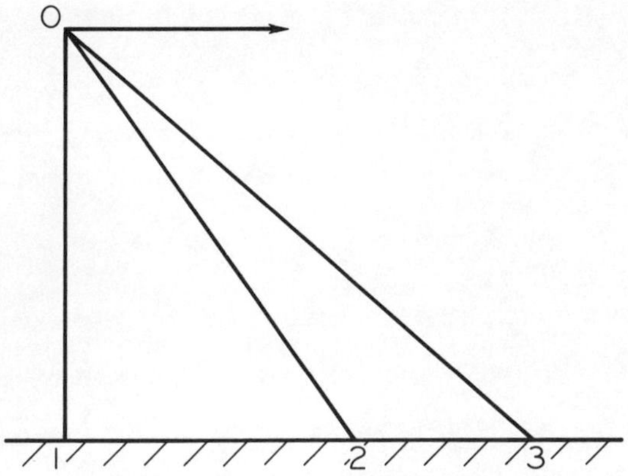

O
1
2
3

(2) Determine the stress in each member of the support frame. Scale: 1mm = 10 N.

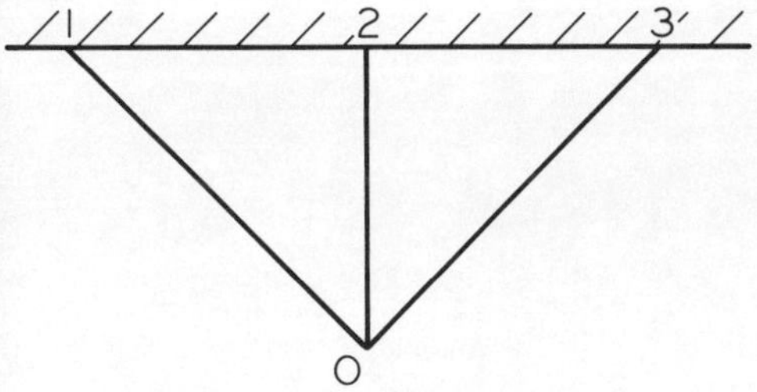

1
2
3
O

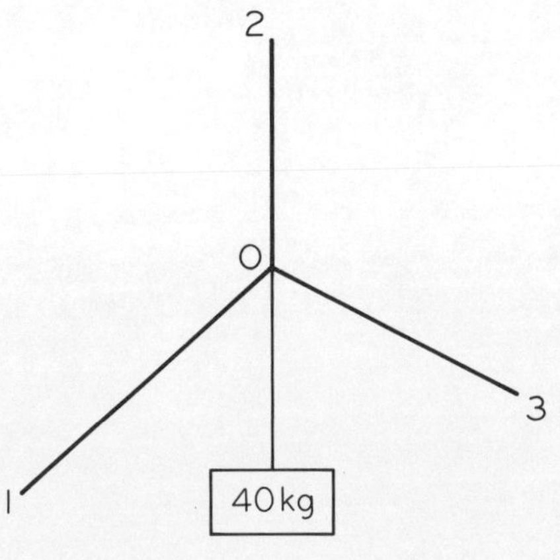

2
O
3
1
40 kg

DRAWN BY

FILE NO.

DRAWI

(1)

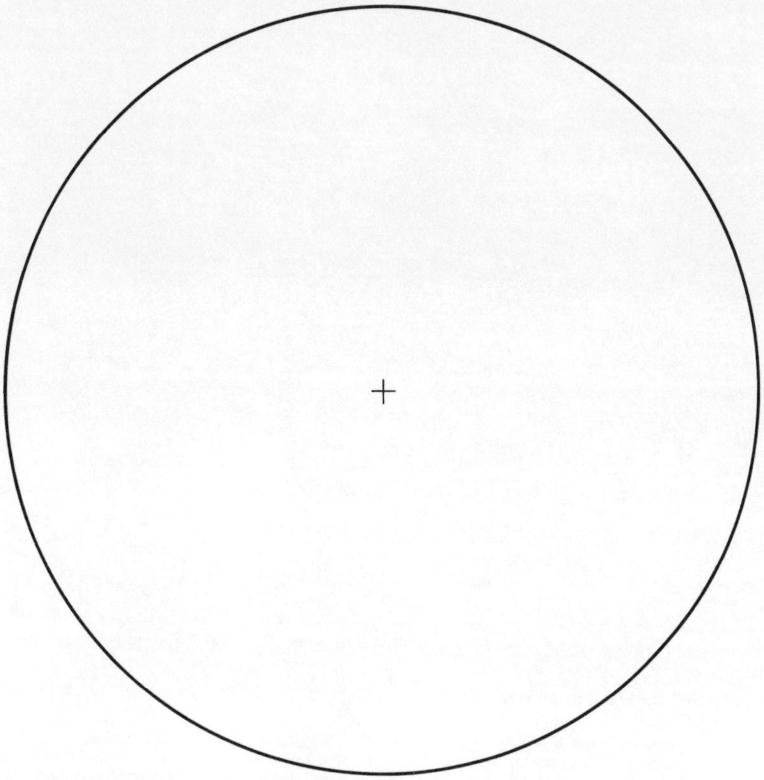

NEW CARS, TRUCKS & BUSES – WORLDWIDE

(2)

400 000
350 000
300 000
250 000
200 000
150 000
100 000
50 000
0
KILOWATT – HOURS (MILLIONS)

PIE AND BAR CHARTS	DRAWN BY	FILE NO.	DRAWING 87

DRAWN BY

FILE NO.

DRAWI

(1)

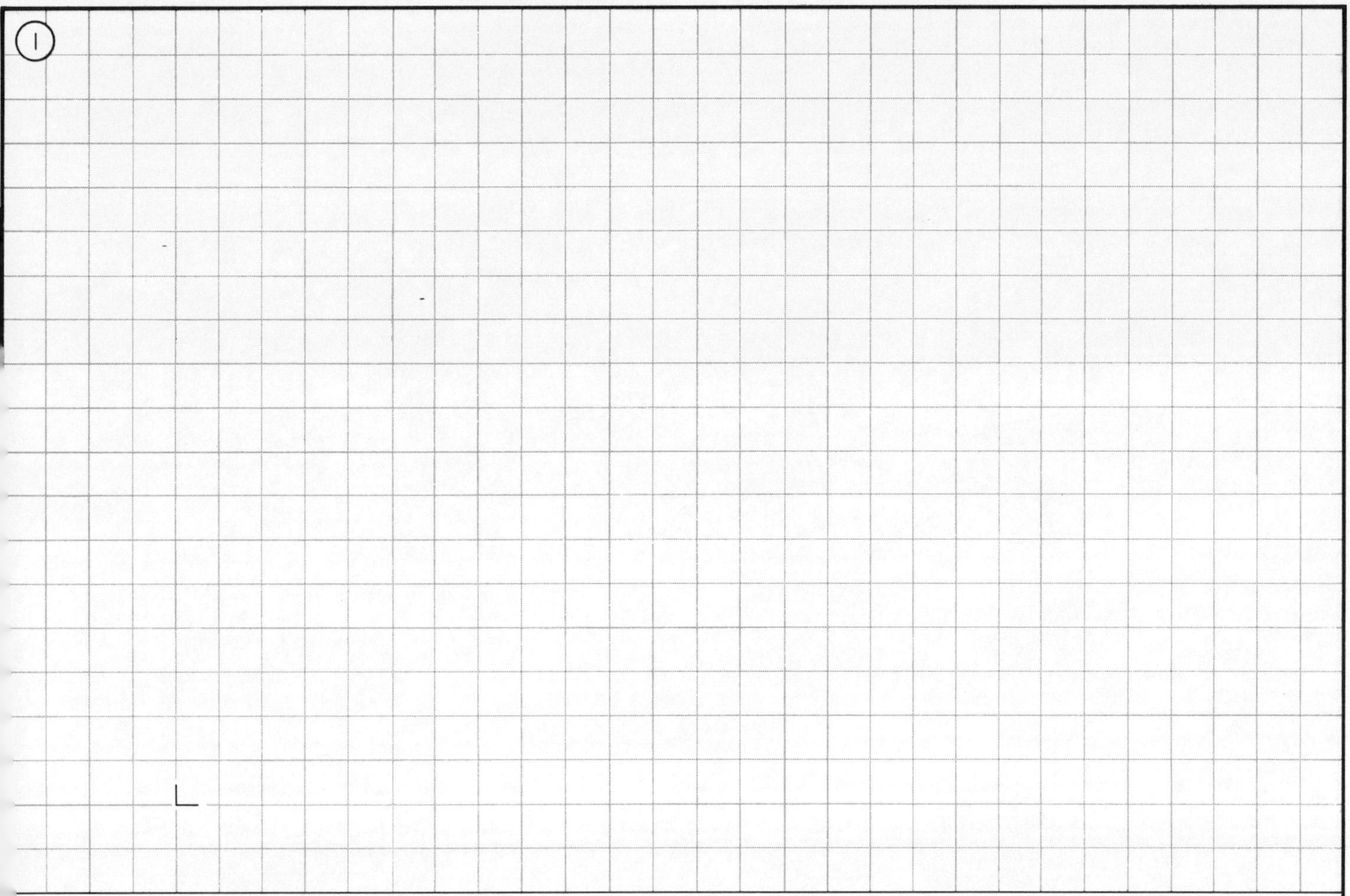

(2)

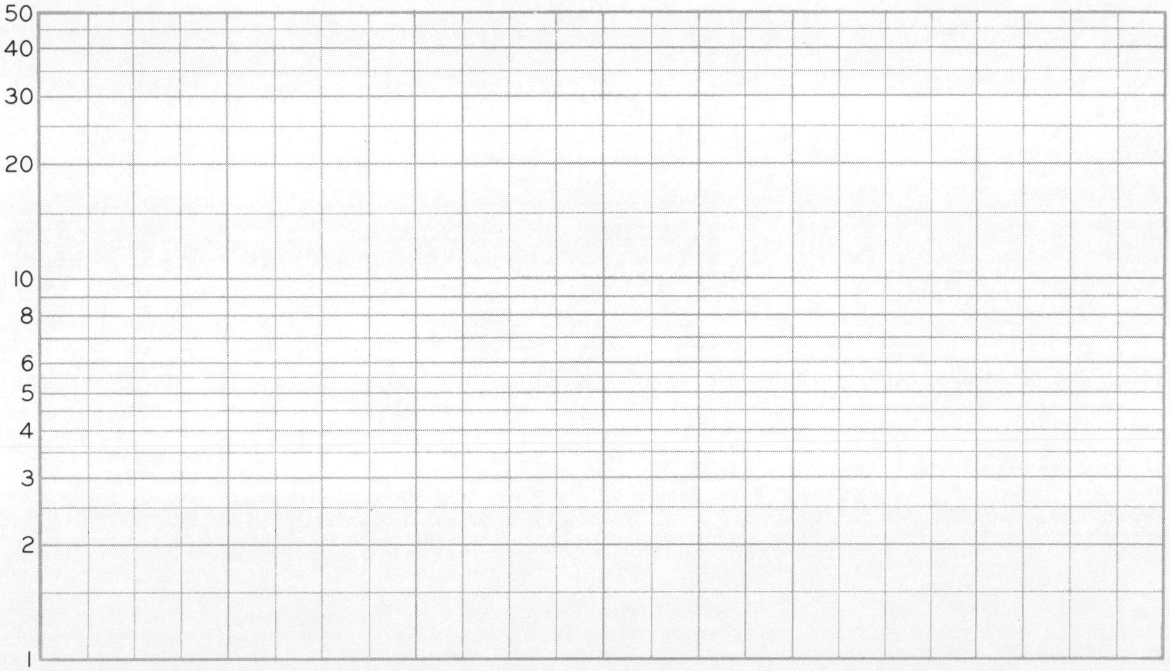
50
40
30
20
10
8
6
5
4
3
2
1

ENGINEERING GRAPHS	DRAWN BY	FILE NO.	DRAWING 88

	DRAWN BY	FILE NO.	DRAW

INITIAL VELOCITY (V_o) – FT/SEC
0
10
20
30
40
TIME (T) – SECS
0
2
4
6
8
10
12
14
16
18
20

(2)

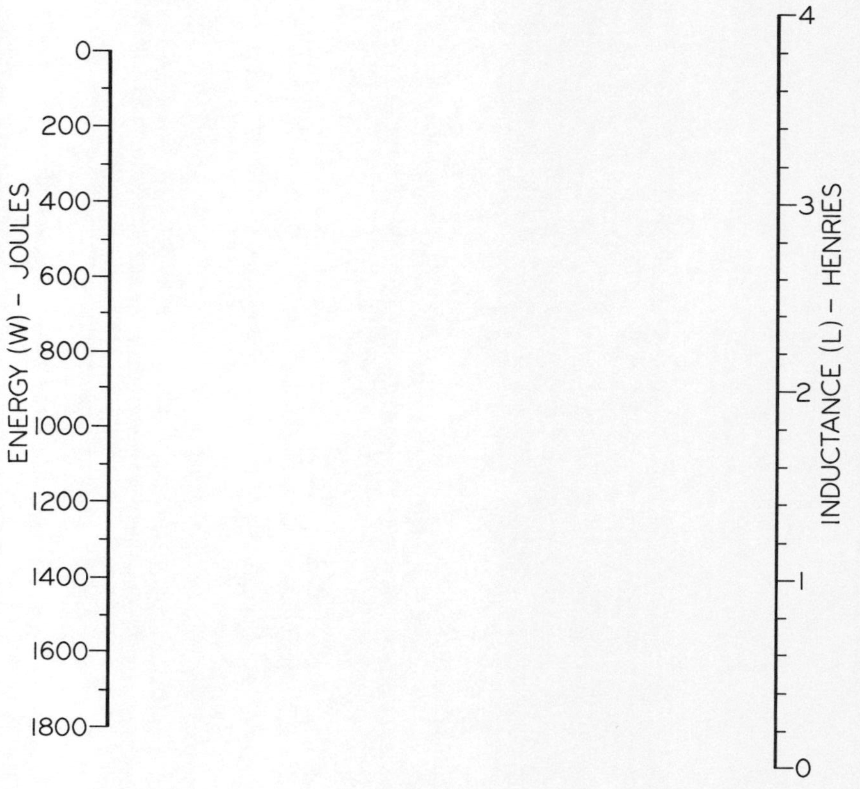
ENERGY (W) – JOULES
0
200
400
600
800
1000
1200
1400
1600
1800
INDUCTANCE (L) – HENRIES
0
1
2
3
4

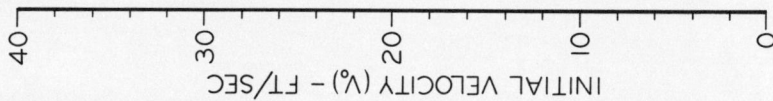

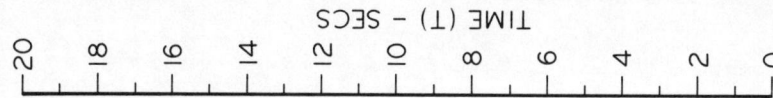

DESIGN TABLE

VAR.	FUNCT.	VARIABLE RANGE	FUNCT. MODULUS	SCALE MODULUS	SCALE LENGTH	DISTANCE FROM CENTER SCALE	DIR.	SCALE USED
V_o	V_o	0 – 40 $\frac{\text{FT.}}{\text{SEC.}}$	$\frac{1}{10}$	$\frac{1}{10}$	4.00"	$\frac{161}{40}$ = 4.025"	↑	10
t	32.2t	0 – 20 SEC.	$\frac{1}{32.2 \times 5} = \frac{1}{161}$	$\frac{1}{5}$	4.00"	$\frac{10}{40}$ = .250"	↑	50
V_f	V_f							

DATA TABLE

I	0	2	4	6	8	10	12	14
VALUES OF W WITH L = 4 ($W = 2I^2$)								

I	16	18	20	22	24	26	28	30

PARALLEL SCALE AND N-CHARTS

ALIGNMENT CHARTS

DRAWN BY

FILE NO.

DRAWING 89

DRAWN BY	FILE NO.	DRAWN

0.6
0.5
0.4
0.3
0.2
COEFFICIENT OF FRICTION (μ)
150
100
80
60
50
40
30
20
15
10
5
NORMAL FORCE (N) - LBS

<u>DESIGN TABLE</u>

VAR.	FUNCTION	m	M	SCALE LENGTH	SCALE DIR.	DISTANCE FROM CENTER SCALE	SCALE USED
μ	LOG μ	10	10	4.77"	↑	4.50"	10
N	LOG N	3.33	3.33	4.96"	↑	1.50"	30
f	LOG f						

LOGARITHMIC PARALLEL-SCALE ALIGNMENT CHART	DRAWN BY	FILE NO.	DRAWING 90

	DRAWN BY	FILE NO.	DRAWI

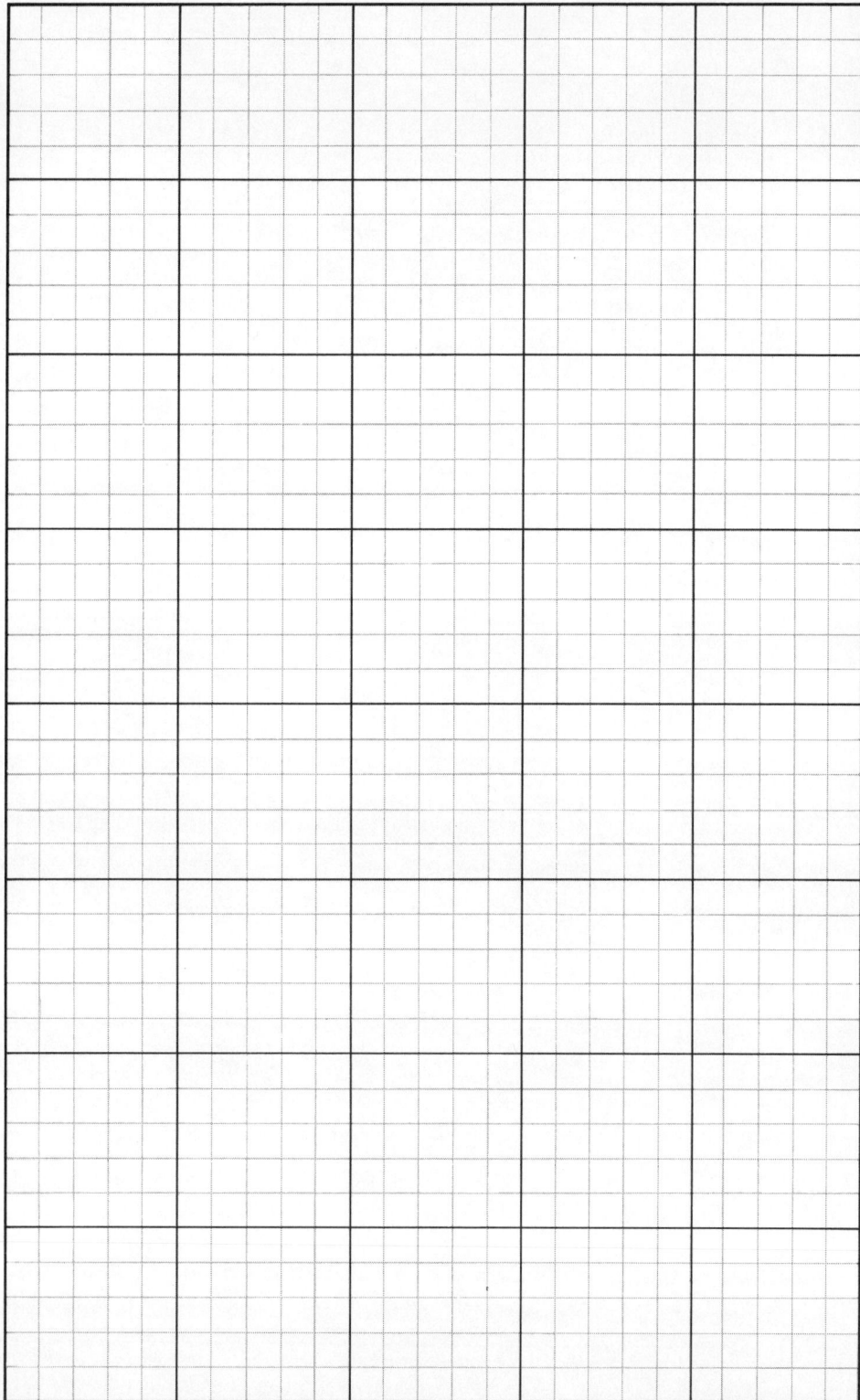

BOTTOM OF DRAWING

RECTANGULAR COORDINATE EMPIRICAL EQUATION	DRAWN BY	FILE NO.	DRAWING 91

	DRAWN BY	FILE NO.	DRAWI

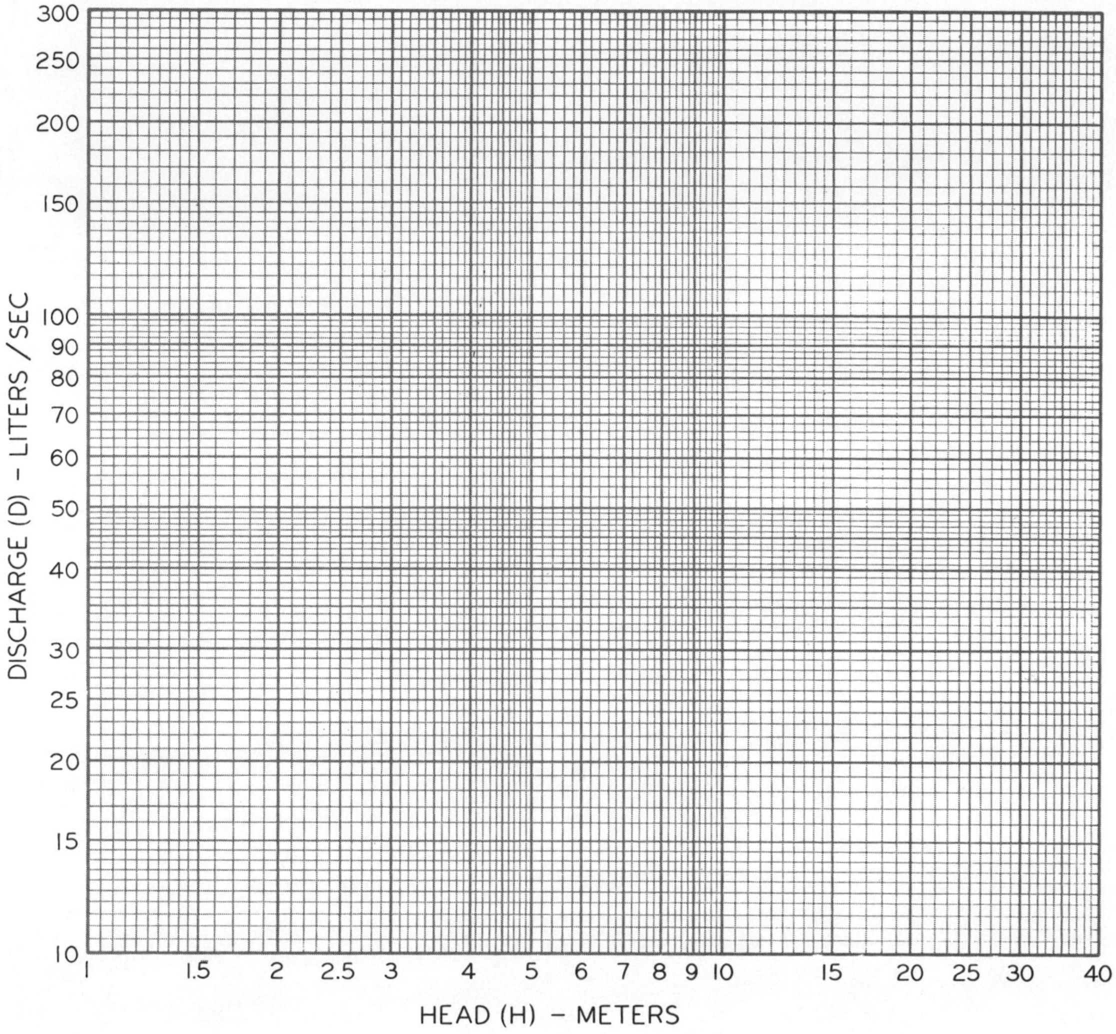
300
250
200
150
100
90
80
70
60
50
40
30
25
20
15
10
DISCHARGE (D) - LITERS / SEC
1
1.5
2
2.5
3
4
5
6
7
8
9
10
15
20
25
30
40
HEAD (H) - METERS

LOG-LOG COORDINATES EMPIRICAL EQUATIONS	DRAWN BY	FILE NO.	DRAWING 92

	DRAWN BY	FILE NO.	DRAW

1. Find the values common to the equations:
$x - 2y - 2 = 0$ and $2y = x^2 - 2x - 14$
2. Find the roots of the equation:
$2y = x^2 - 2x - 14$

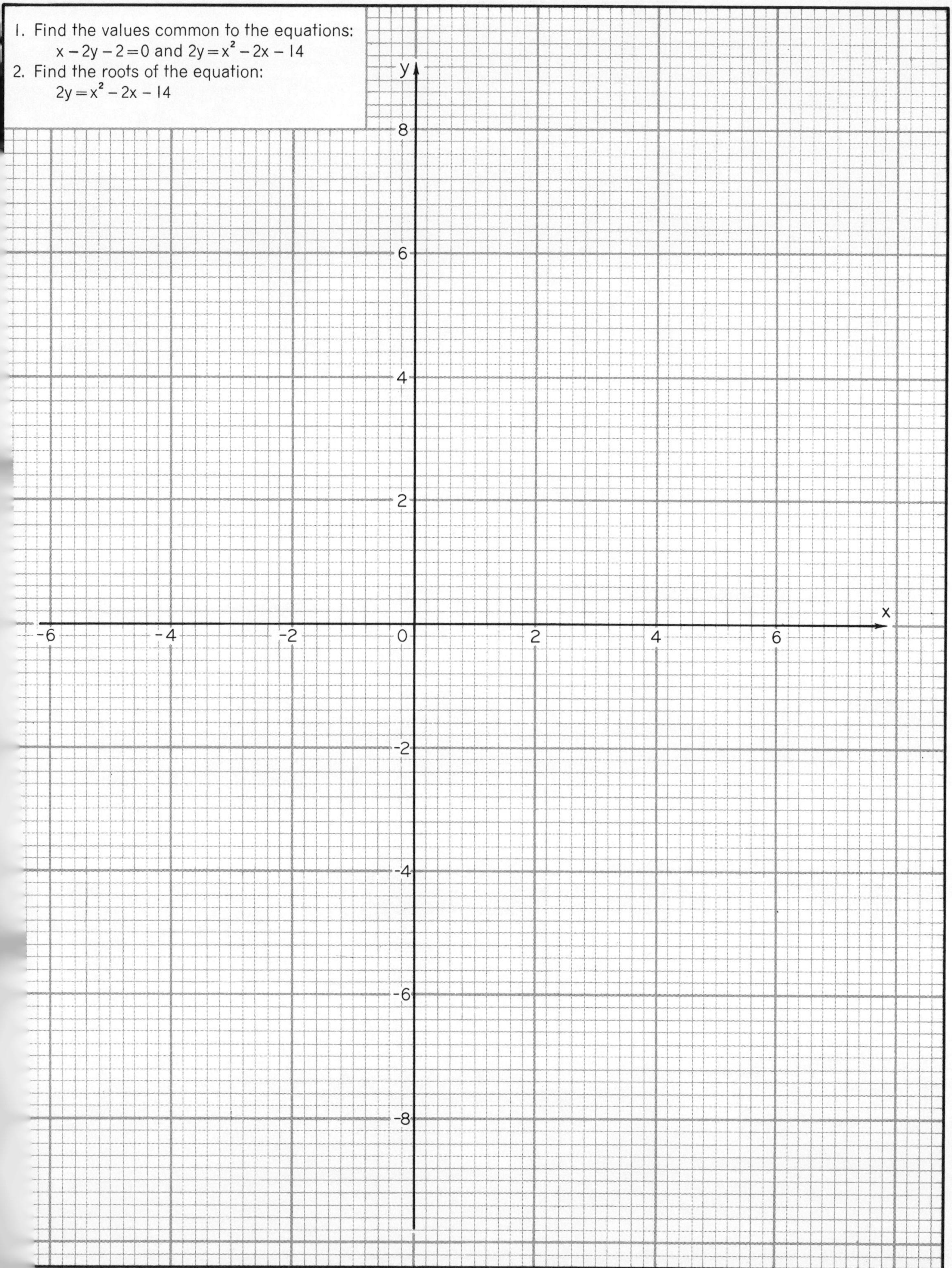

y
8
6
4
2
0
-2
-4
-6
-8
x
-6
-4
-2
2
4
6

ALGEBRAIC SOLUTIONS GRAPHICAL MATHEMATICS	DRAWN BY	FILE NO.	DRAWING 93

	DRAWN BY	FILE NO.	DRAW

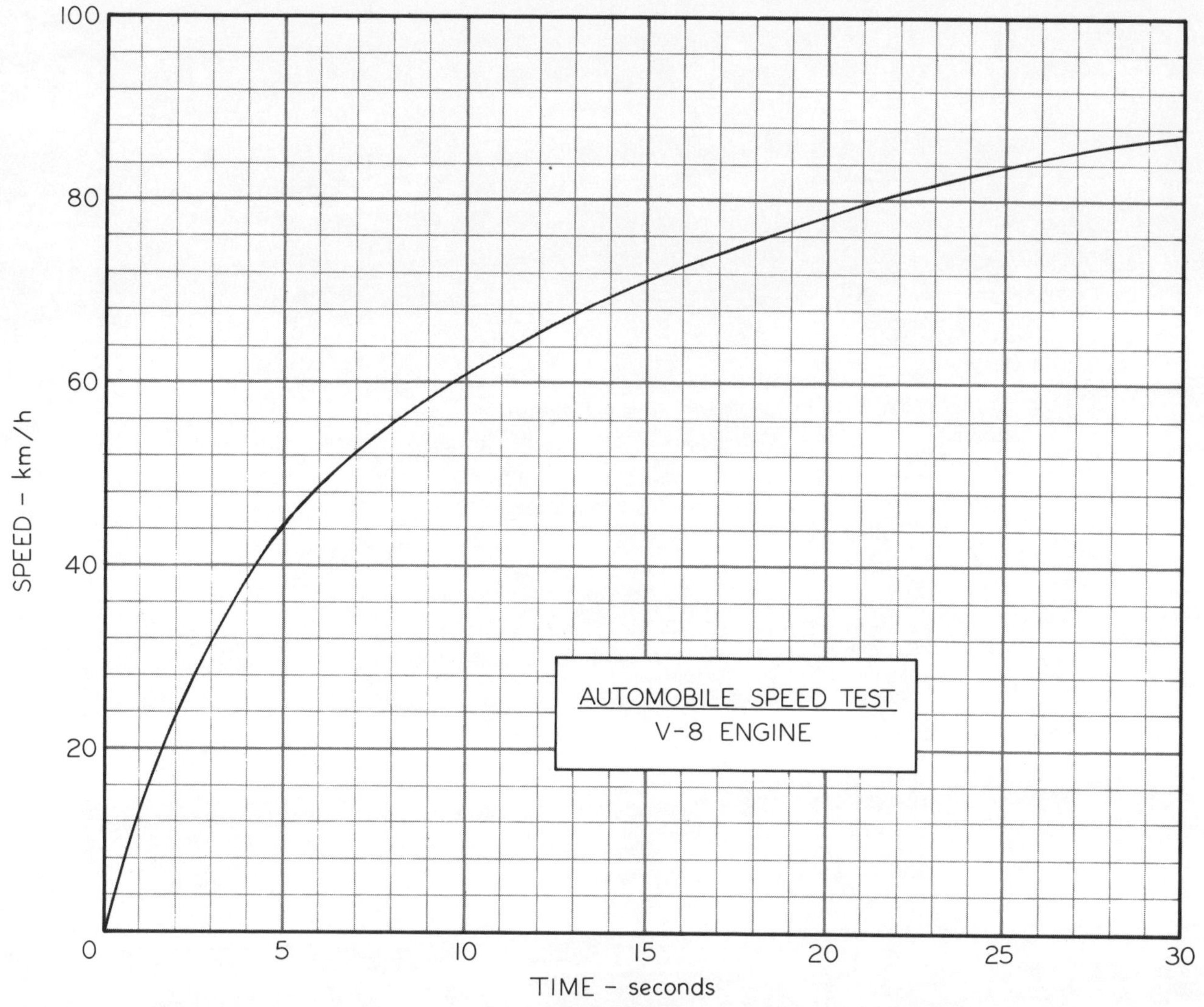

AUTOMOBILE SPEED TEST
V-8 ENGINE
SPEED – km/h
100
80
60
40
20
0
5
10
15
20
25
30
TIME – seconds

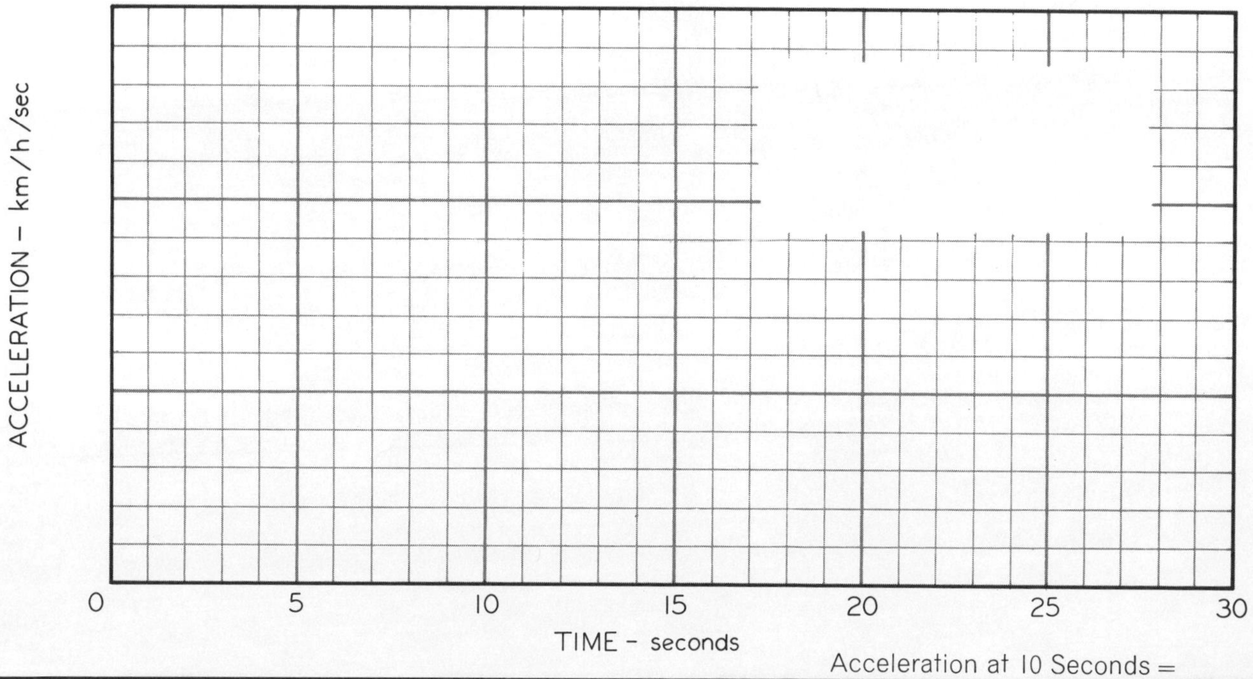

ACCELERATION – km/h/sec
0
5
10
15
20
25
30
TIME – seconds

Acceleration at 10 Seconds =

DIFFERENTIATION GRAPHICAL MATHEMATICS	DRAWN BY	FILE NO.	DRAWING 94

DRAWN BY

FILE NO.

DRAWIN

FUEL TANK — AREA

ONE-FOURTH SHOWN

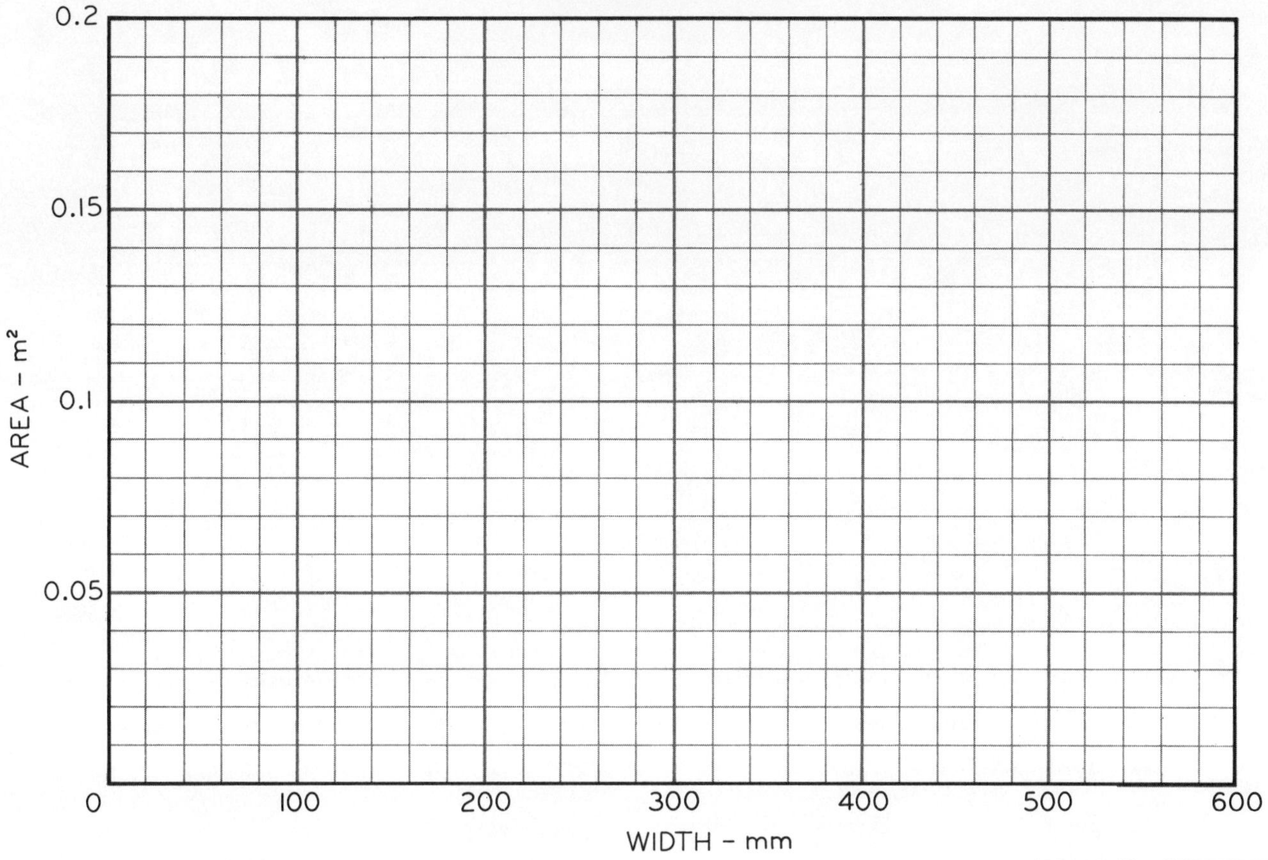
AREA – m^2
0.2
0.15
0.1
0.05
0
100
200
300
400
500
600
WIDTH – mm

VOLUME = ____________

FUEL TANK — CROSS SECTION

ONE-FOURTH SHOWN

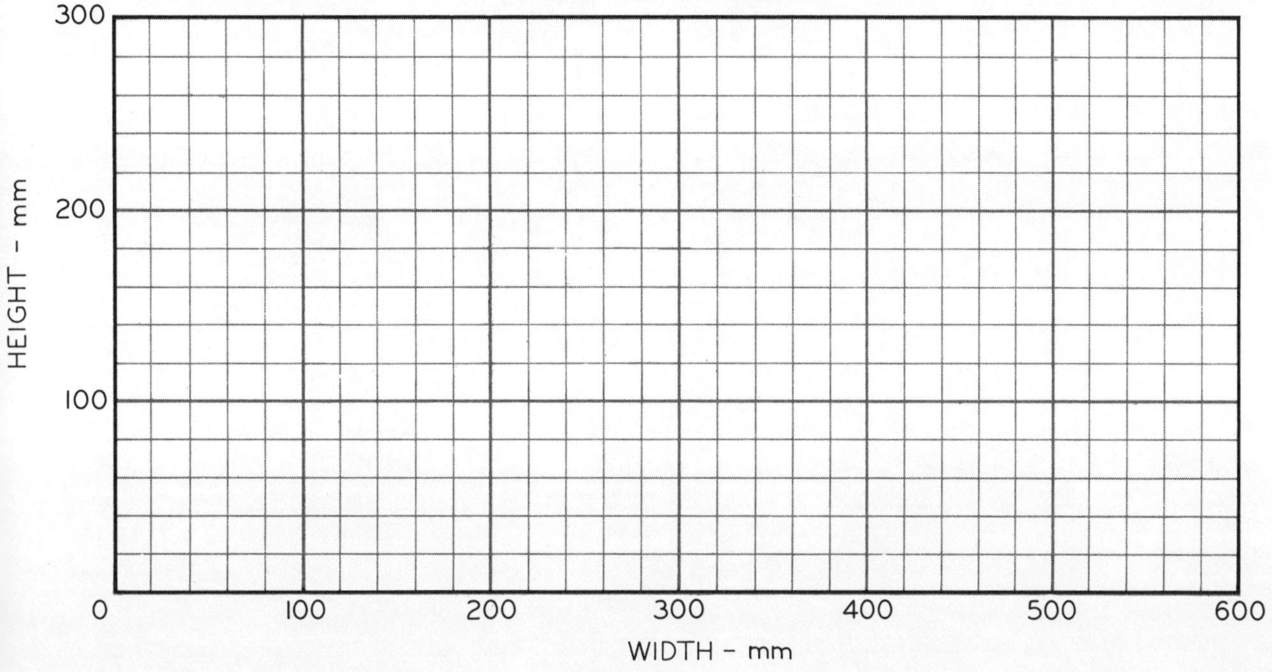
HEIGHT – mm
300
200
100
0
100
200
300
400
500
600
WIDTH – mm

E–

INTEGRATION GRAPHICAL MATHEMATICS	DRAWN BY	FILE NO.	DRAWING 95

DRAWN BY | FILE NO. | DRAWI

Complete the table of TERMS by entering the letter identifiers of the matching descriptions.

TERMS

CURSOR	
DIGITIZER TABLET	
GRAPHIC PRIMITIVE	
PIXEL	
RESOLUTION	
RASTER DISPLAY	
RAM	
DEBUG	
HARD COPY	
ANALOG	
DIGITAL	
CAE	
COMMAND	
PLOTTER	
HARD DISK	
VECTOR	
COORDINATE SYSTEM	
MENU	
BIT	
MOUSE	
SOFTWARE	
JOY STICK	
WINDOW	
TRANSFORM	
BYTE	
LIGHTPEN	

Descriptions

A Handheld pointing device for pick and coordinate entry

B Computer program to perform specific tasks

C Counts in discrete steps or digits

D Smallest unit of digital information

E Collection of commands for selection

F Device to convert analog picture to coordinate digital data

G Fundamental drawing entity

H Picture element dot in a display grid

I Random Access Memory - volatile physical memory

J Continuous measurements without steps

K Computer assisted engineering

L Group of 8 bits commonly used to represent a character

M Paper printout

N Hand controlled lever used as input device

O Smallest spacing between CRT display elements

P Convert an image into a proper display format

Q Directed line segment with magnitude

R Flicker-free scanned CRT surface

S A bounded rectangular area on screen

T A visual tracking symbol

U Handheld photosensitive input device

V Control signal

W Correct errors

X Non-volatile external storage device

Y Hard copy device for vector drawing

Z Common reference system for spatial relationships

DRAWN BY

FILE NO.

DRAWIN

① Complete the table by defining X and Y coordinates of the given points .

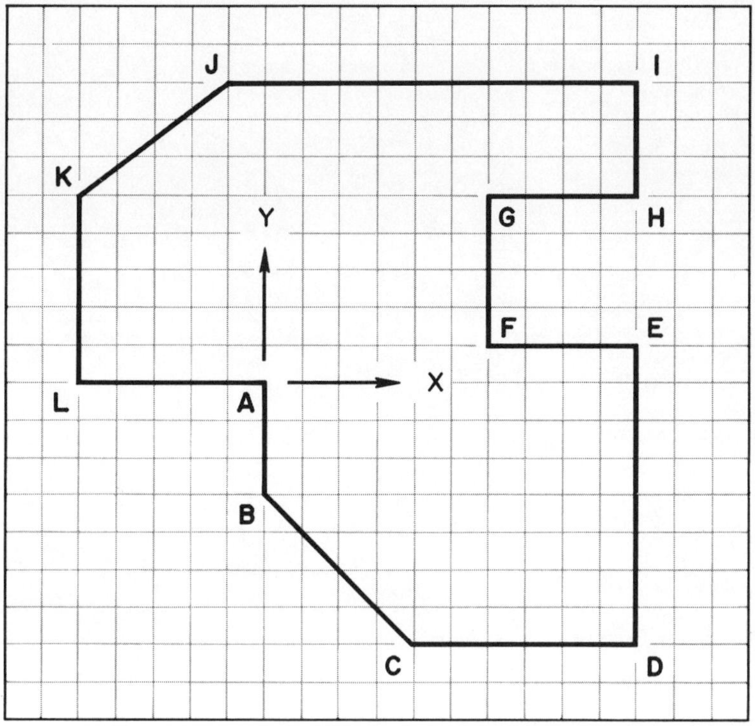
J
I
K
Y
G
H
F
E
L
A
X
B
C
D

Point	Coordinate	
	X	Y
A		
B		
C		
D		
E		
F		
G		
H		
I		
J		
K		
L		

② Plot the given points on the grid and draw the view .

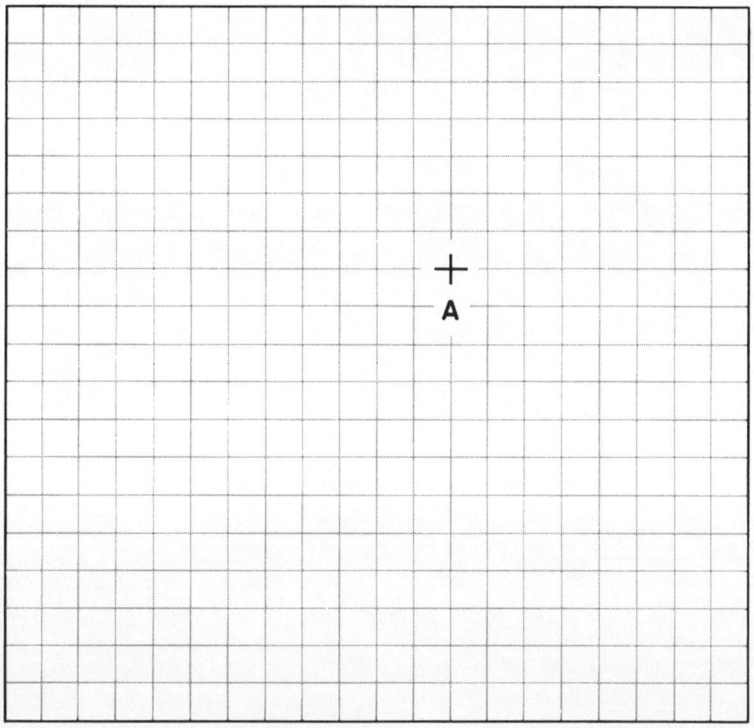
A

Point	Coordinate	
	X	Y
A	0	0
B	0	4
C	−5	4
D	−5	−2
E	−10	−6
F	−10	−10
G	−2	−10
H	6	−4
I	6	3
J	3	3
K	3	−2
L	0	−4

	DRAWN BY	FILE NO.	DRAW

① Complete the table for drawing the object .

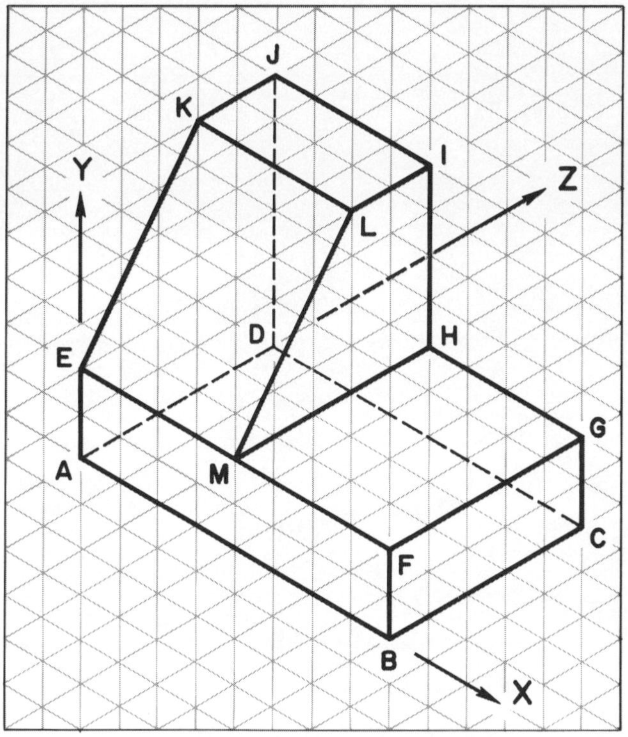
J
K
I
Y
Z
L
D
H
E
G
A
M
C
F
B
X

Point	Move 0 Draw 1	X	Y	Z	Point	Move 0 Draw 1	X	Y	Z
A	0	0	0	0					
B									
C									

② Draw the object based on the data given in the table .

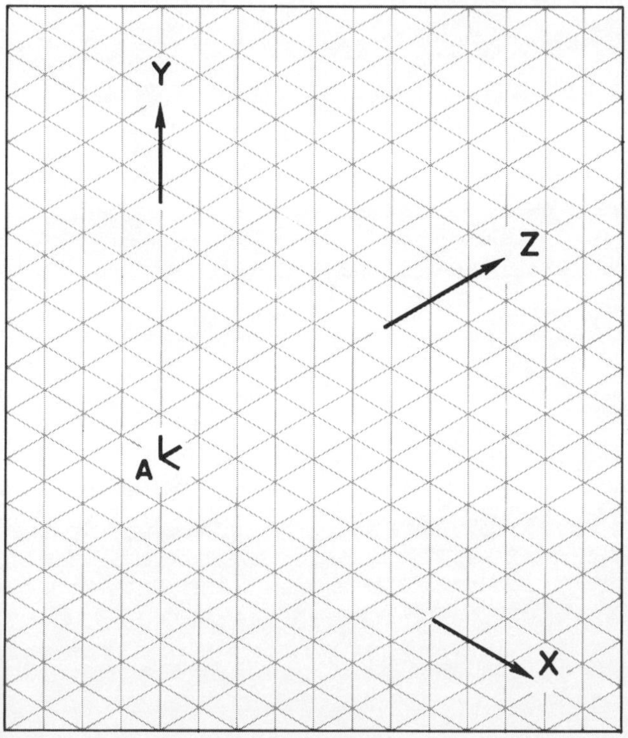
Y
Z
A
X

Point	Move 0 Draw 1	X	Y	Z	Point	Move 0 Draw 1	X	Y	Z
A	0	0	0	0	G	1	−3	−4	0
B	1	0	5	0	J	1	−3	−4	5
C	1	0	5	5	L	0	−3	−1	0
D	1	0	0	5	B	1	0	5	0
E	1	6	0	5	K	0	−3	−1	5
F	1	6	0	0	C	1	0	5	5
A	1	0	0	0	H	0	0	−4	0
D	1	0	0	5	F	1	6	0	0
G	0	−3	−4	0	I	0	0	−4	5
L	1	−3	−1	0	E	1	6	0	5
K	1	−3	−1	5					
J	1	−3	−4	5					
I	1	0	−4	5					
H	1	0	−4	0					

	DRAWN BY	FILE NO.	DRAW

Complete the table by entering the Menu Selections used for generating the drawing

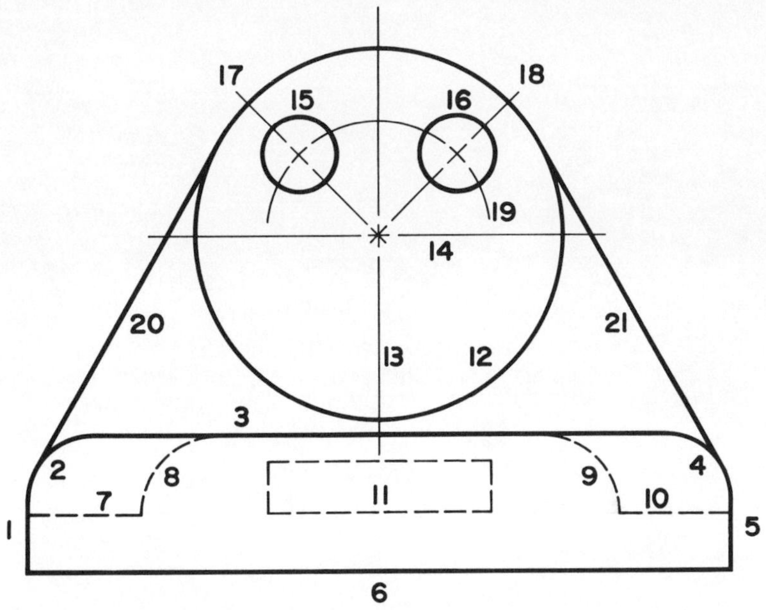
17
18
15
16
19
14
20
21
13
12
3
2
4
7
8
11
9
10
1
5
6

Entity	Line type menu selection	Entity menu selection	Construction menu selection
1			
2			
3			
4			
5			
6			
7			
8			
9			
10			
11			
12			
13			
14			
15			
16			
17			
18			
19			
20			
21			

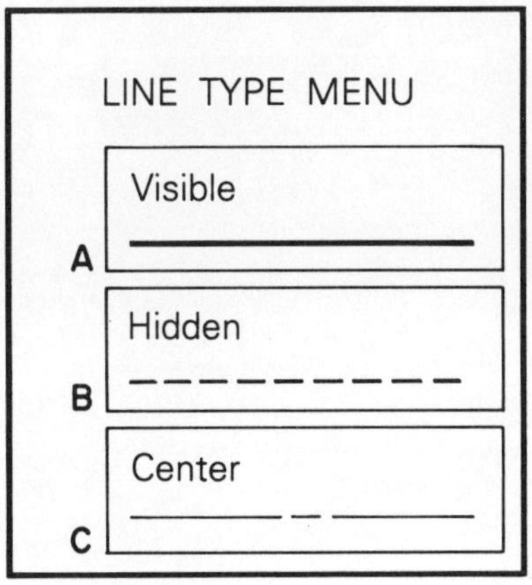
LINE TYPE MENU
A Visible
B Hidden
C Center

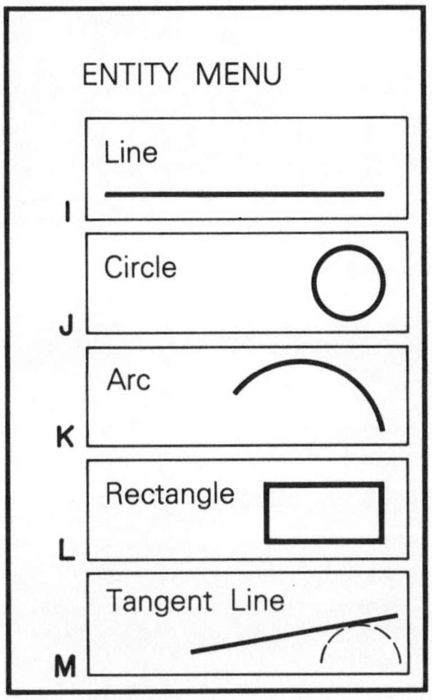
ENTITY MENU
I Line
J Circle
K Arc
L Rectangle
M Tangent Line

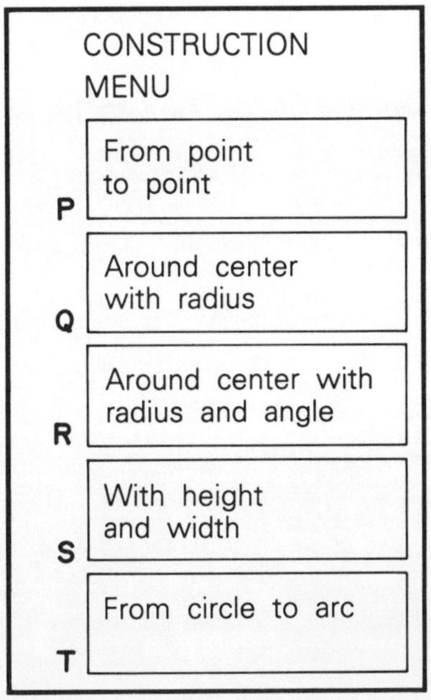
CONSTRUCTION MENU
P From point to point
Q Around center with radius
R Around center with radius and angle
S With height and width
T From circle to arc

	DRAWN BY	FILE NO.	DRAWI

Description of VIEW COORDINATES

VIEW COORDINATES are the coordinate values of the object as assigned with respect to the computer screen, with X , Y and Z axes positioned as shown below . The coordinates remain the same irrespective of the view selected on the screen .

Axis	Position	Positive Direction
X	Horizontal	To the right
Y	Vertical	Toward the top
Z	Perpendicular to the screen	Outward from the screen

Description of WORLD COORDINATES

WORLD COORDINATES are the coordinate values of the object as assigned with respect to the axes of the object . The X , Y and Z axes are positioned as shown, such that for the top view the X axis is horizontal to the right, the Y axis is vertical to the top and the Z axis is perpendicular to the screen positioned outwards . The coordinates in relation to the screen change according to the view selected on the screen.

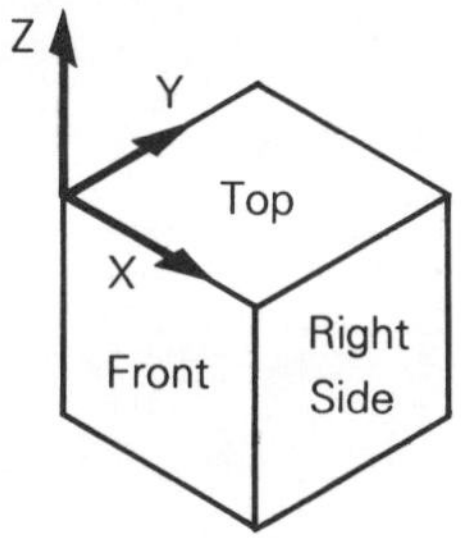

Z
Y
X
Top
Front
Right
Side

Complete the tables by entering the VIEW and WORLD COORDINATES of the given points of the object .

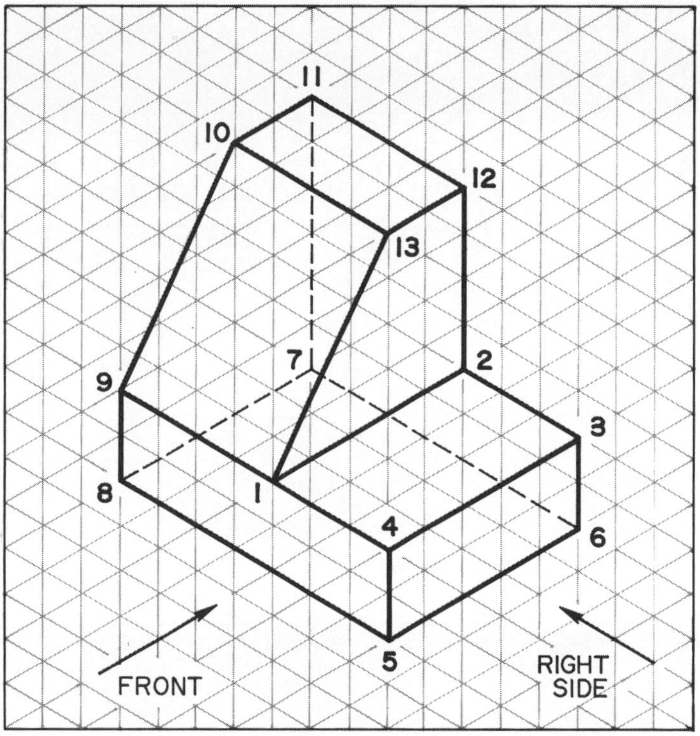

11
10
12
13
7
2
9
3
8
1
4
6
5
FRONT
RIGHT
SIDE

FRONT VIEW						
Points	View Coordinates			World Coordinates		
	X	Y	Z	X	Y	Z
1						
2						
3						
4						
5						
6						
7						
8						
9						
10						
11						
12						
13						

RIGHT SIDE VIEW						
Points	View Coordinates			World Coordinates		
	X	Y	Z	X	Y	Z
1						
2						
3						
4						
5						
6						
7						
8						
9						
10						
11						
12						
13						

	DRAWN BY	FILE NO.	DRAW

1 CENTERING BRACKET

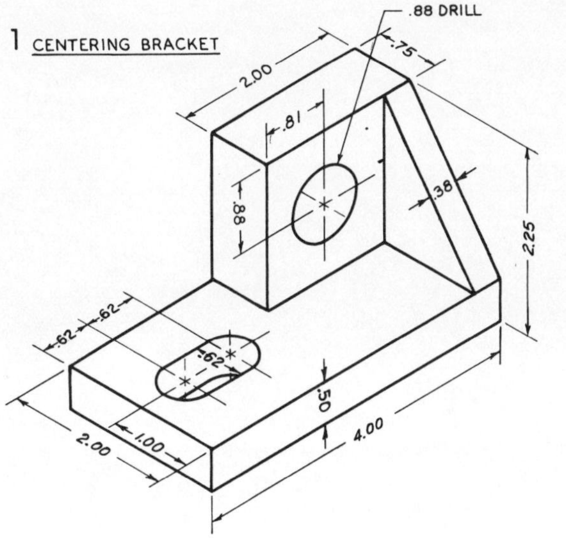

.88 DRILL
2.00
.75
.81
.88
.38
2.25
.62
.62
.62
.50
2.00
1.00
4.00

2 HINGE

METRIC

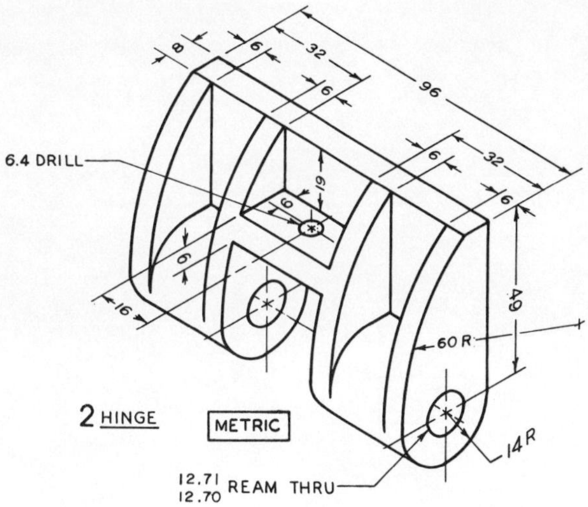

8
6
32
6
96
6.4 DRILL
6
32
6
6
19
6
16
49
60 R
14R
12.71
12.70 REAM THRU

3 GRIPPING JAW

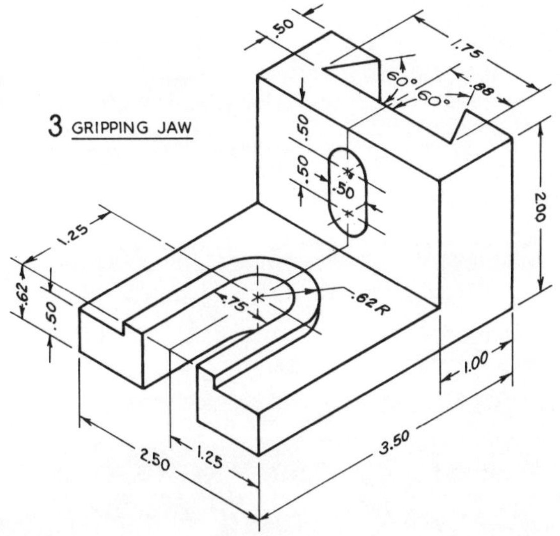

.50
1.75
60°
60°
.88
.50
.50
.50
2.00
1.25
.62
.50
.75
.62R
1.00
2.50
1.25
3.50

4 GUIDE BLOCK

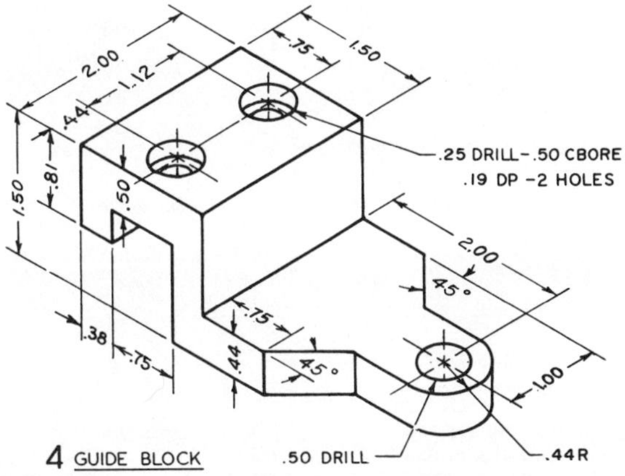

2.00
1.12
.75
1.50
.44
.81
1.50
.50
.25 DRILL-.50 CBORE
.19 DP -2 HOLES
2.00
45°
.75
.38
.75
.44
45°
1.00
.50 DRILL
.44R

5 HINGE BASE

METRIC

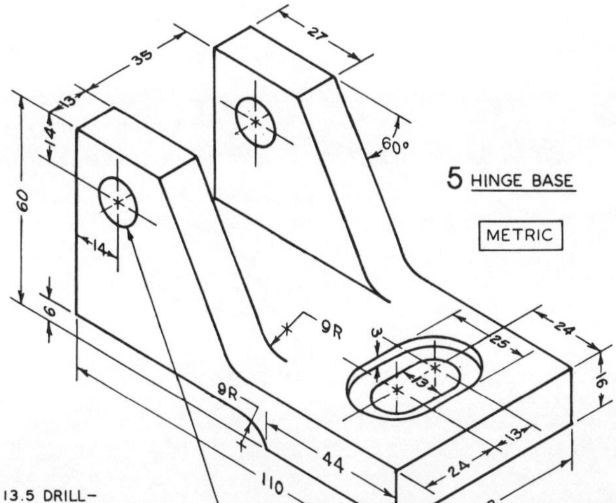

35
27
13
14
60°
60
14
6
9R
3
25
24
13
16
9R
44
24
13
110
60
13.5 DRILL-
2 HOLES IN LINE

6 PIVOT PLATE

METRIC

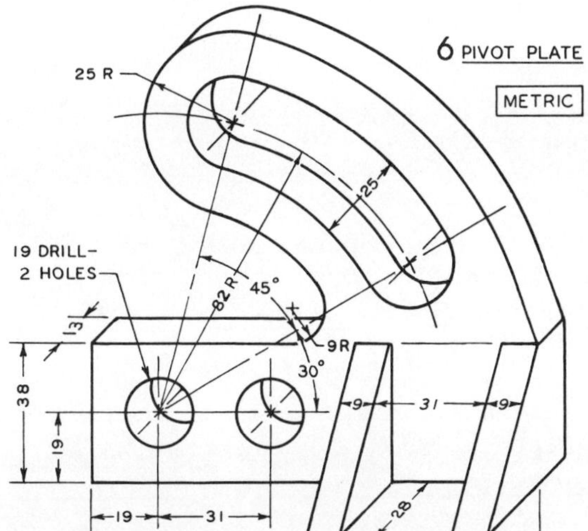

25 R
25
19 DRILL-
2 HOLES
82 R
45°
9R
13
30°
38
9
31
9
19
19
31
28
127

DETAIL DRAWINGS	DRAW OR SKETCH THE NECESSARY VIEWS OF THE OBJECT ASSIGNED. DIMENSION COMPLETELY	FILE NO.	DRAWING 101

	DRAWN BY	FILE NO.	DRAWI

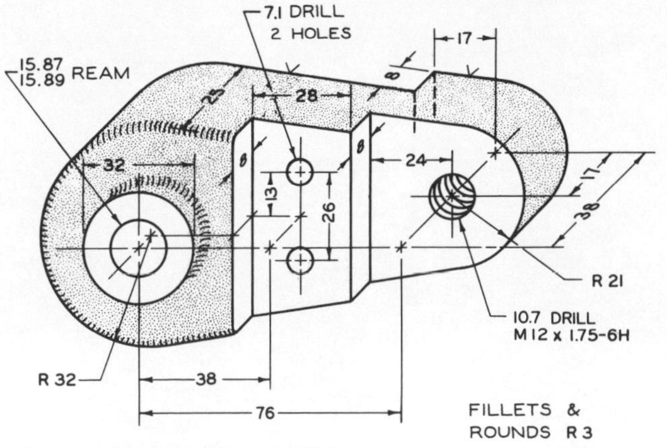
7.1 DRILL
2 HOLES
15.87
15.89 REAM
17
8
28
25
32
8
13
26
8
24
17
38
R 21
10.7 DRILL
M12 x 1.75-6H
R 32
38
76
FILLETS &
ROUNDS R3

1 SPACER ARM
C1—2 REQD

METRIC

2 DRIVE TIGHTENER BASE
C1—1 REQD

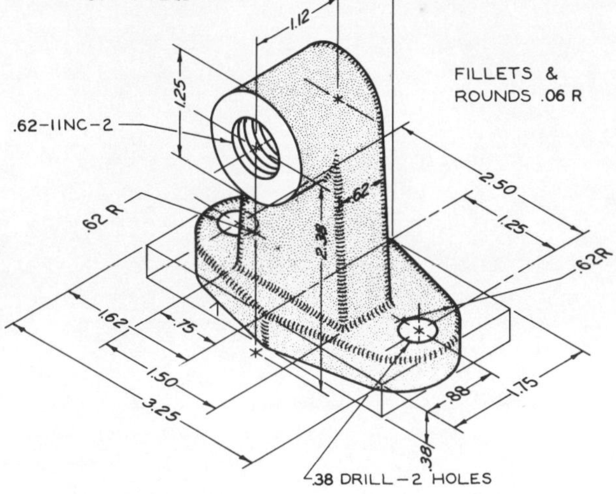
.75
1.12
1.25
FILLETS &
ROUNDS .06 R
.62-11NC-2
2.50
.62
.62 R
2.38
1.25
.62R
1.62
.75
1.50
3.25
.88
1.75
.38
.38 DRILL-2 HOLES

3 BASE ANGLE FLANGE
C1—1 REQD

METRIC

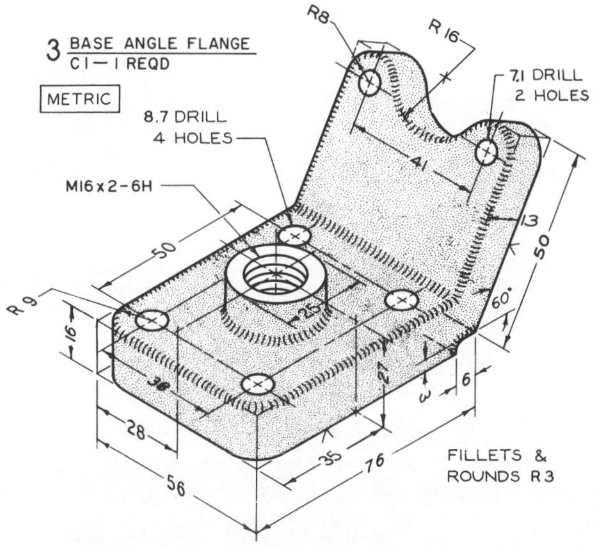
R8
R 16
7.1 DRILL
2 HOLES
8.7 DRILL
4 HOLES
M16 x 2-6H
41
13
50
50
60°
R 9
16
25
38
27
3
6
28
35
76
56
FILLETS &
ROUNDS R3

4 CENTER GUIDE
CRS—2 REQD

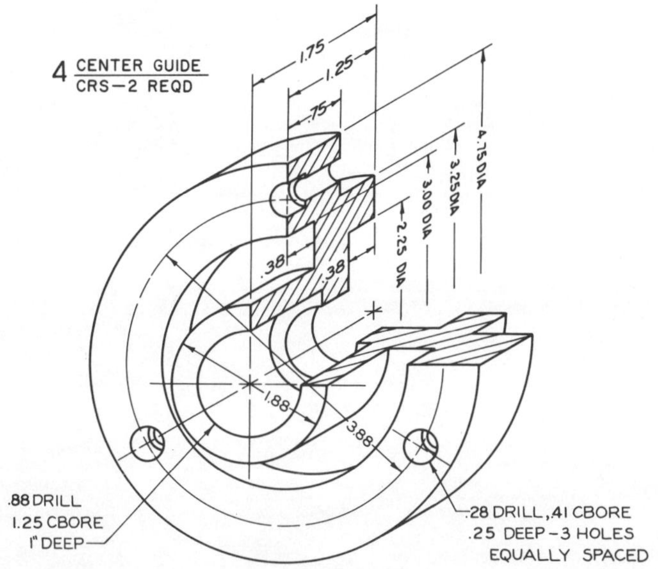
1.75
1.25
.75
4.75 DIA
3.25 DIA
3.00 DIA
2.25 DIA
.38
.38
1.88
3.88
.88 DRILL
1.25 CBORE
1" DEEP
.28 DRILL, .41 CBORE
.25 DEEP-3 HOLES
EQUALLY SPACED

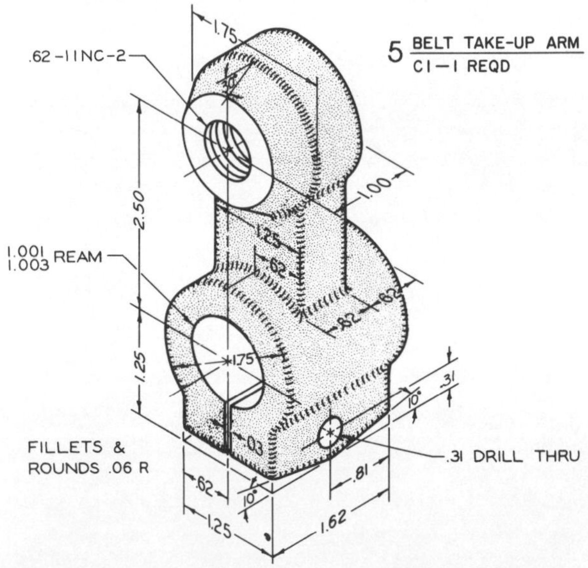
1.75
.62-11NC-2
30°
1.00
2.50
1.001
1.003 REAM
.25
.62
.82
.62
1.25
1.75
10°
.31
.03
.31 DRILL THRU
FILLETS &
ROUNDS .06 R
.62
.81
10°
1.25
1.62

5 BELT TAKE-UP ARM
C1—1 REQD

6 TRIP BRACKET
C1—1 REQD

METRIC

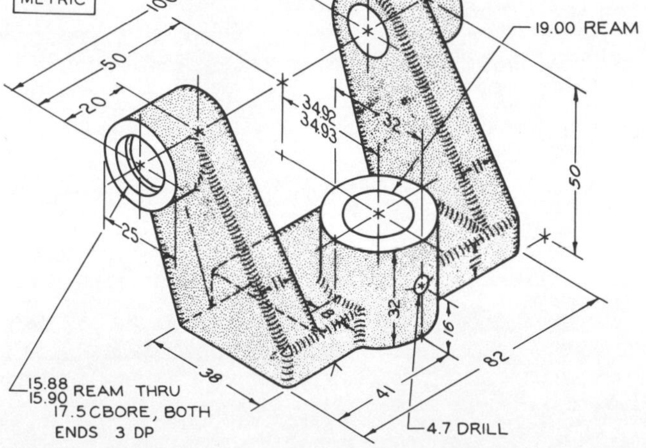
FILLETS AND
ROUNDS R3
20
100
19.00 REAM
50
20
34.92
34.93
32
50
25
32
16
6
38
41
82
15.88
15.90 REAM THRU
17.5 CBORE, BOTH
ENDS 3 DP
4.7 DRILL

DETAIL DRAWINGS	DRAW OR SKETCH THE NECESSARY VIEWS OF THE OBJECT ASSIGNED. DIMENSION COMPLETELY	FILE NO.	DRAWING 102

	DRAWN BY	FILE NO.	DRAWIN

	DRAWN BY	FILE NO.	DRAWING

	DRAWN BY	FILE NO.	DRAW

	DRAWN BY	FILE NO.	DRAWING

DRAWN BY

FILE NO.

DRAWI

	DRAWN BY	FILE NO.	DRAWING

	DRAWN BY	FILE NO.	DRAWI

	DRAWN BY	FILE NO.	DRAWING

	DRAWN BY	FILE NO.	DRAW

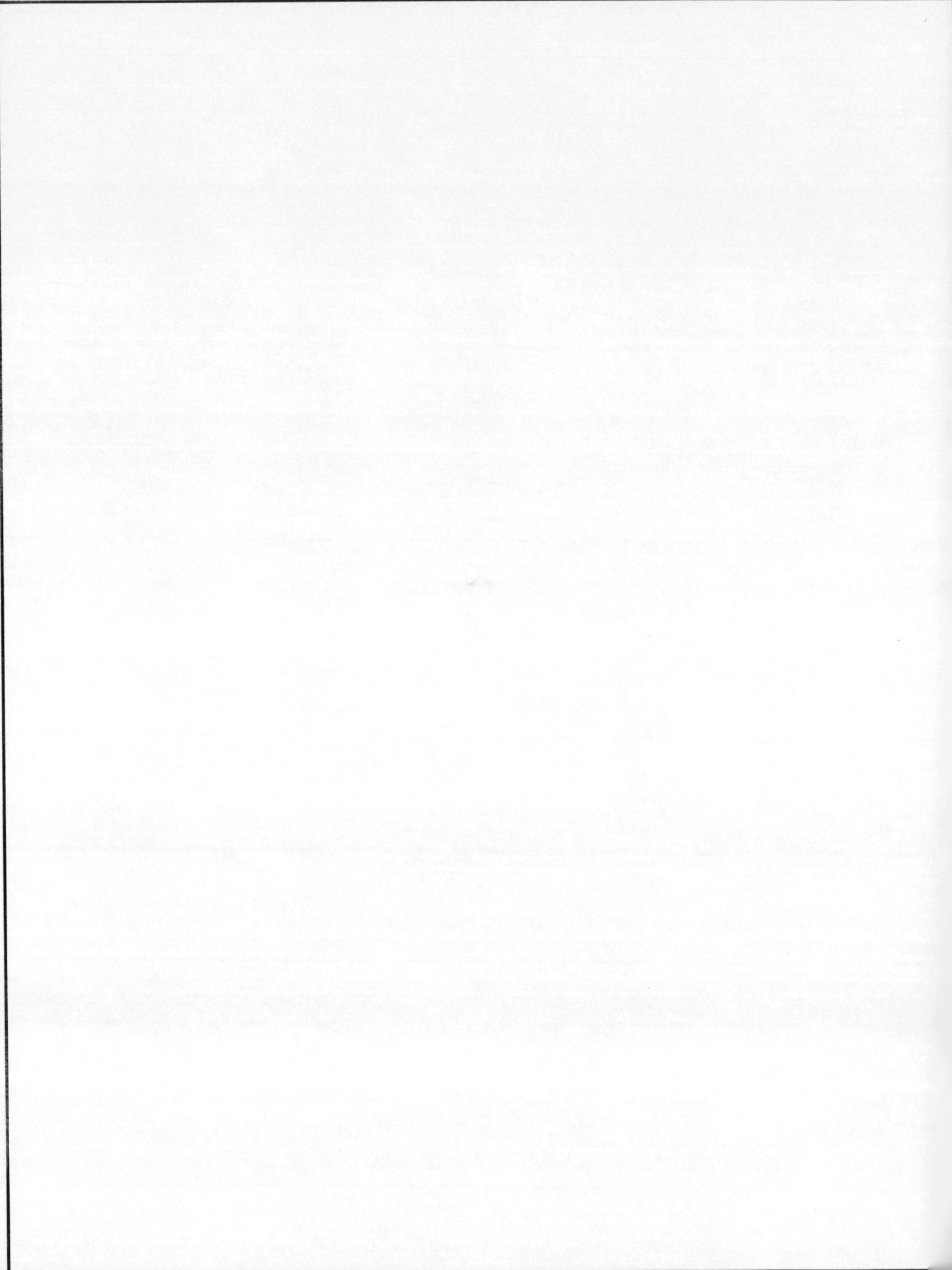

1

SWITCH BLOCK

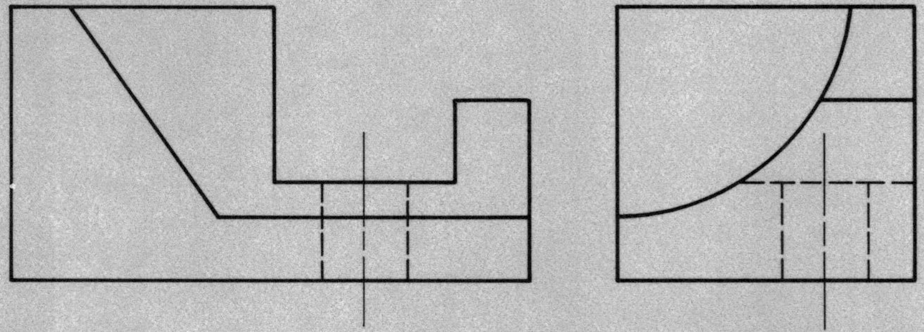

2

RETURN FINGER

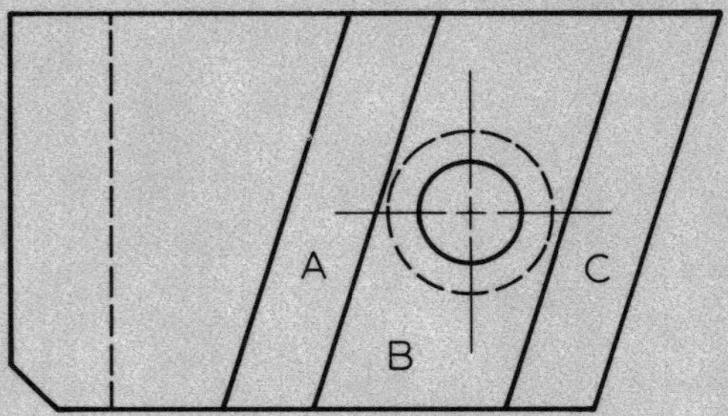
A
B
C

Surfaces A & B are normal surfaces.
Surface C is an oblique surface.

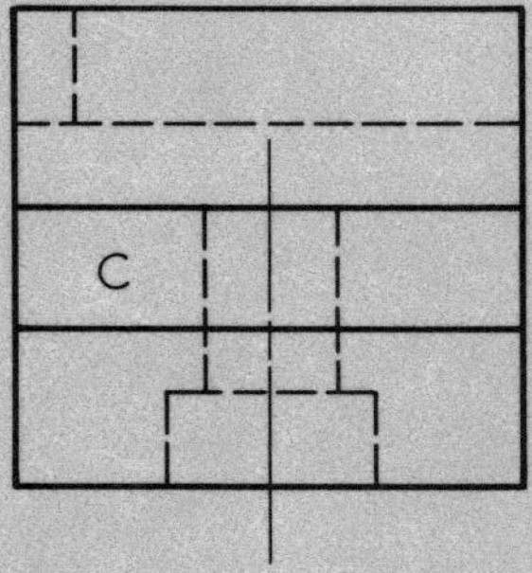
C

3

OSCILLATING ARM

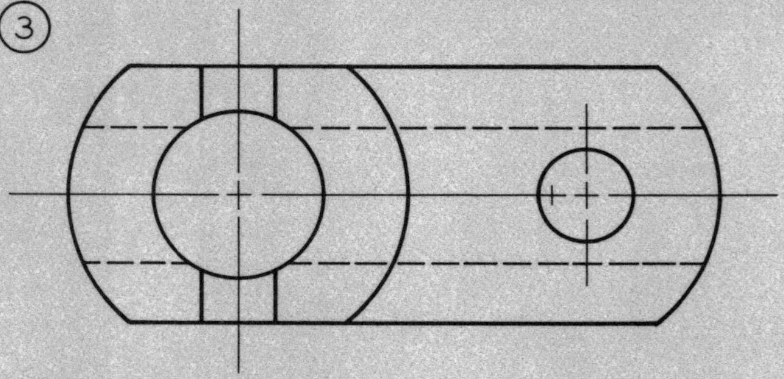

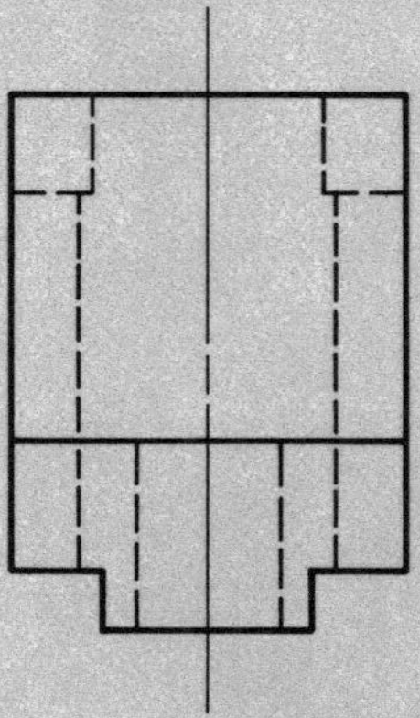

4

INDICATER ROLLER

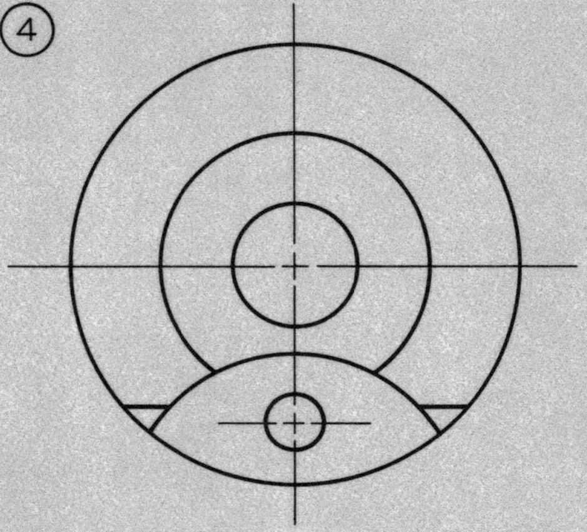

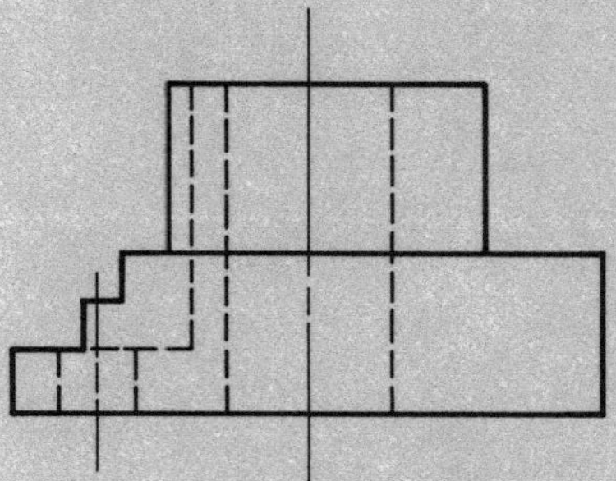

MISSING VIEWS MULTIVIEW PROJECTION	DRAWN BY	FILE NO.	DRAWING 19

MULTIVIEW PROJECTION

MISSING VIEWS

DRAWN BY

FILE NO.

DRAWING 19

(1)

SWITCH BLOCK

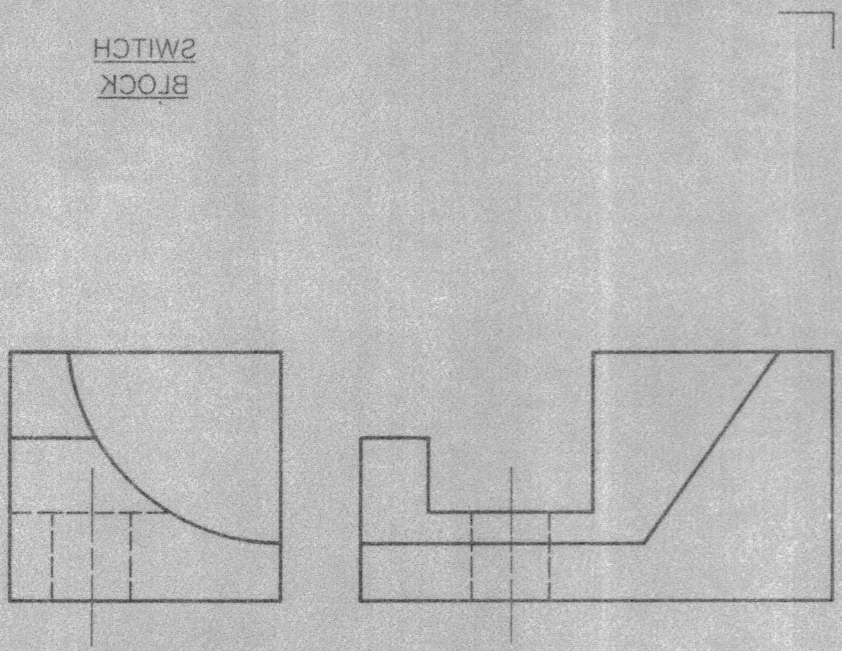

(2)

RETURN FINGER

Surfaces A&B are normal surfaces.
Surface C is an oblique surface.

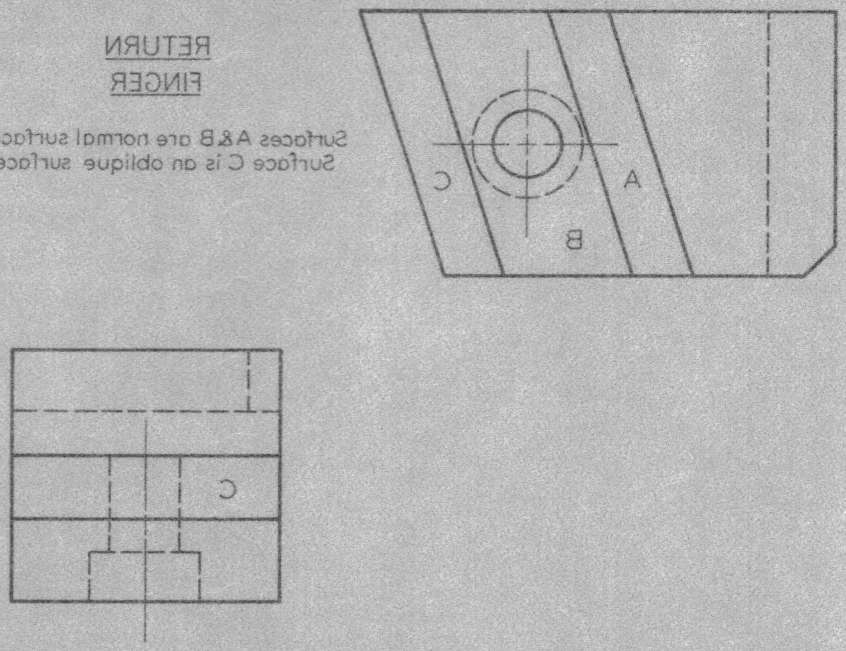

A
B
C
C

(3)

OSCILLATING ARM

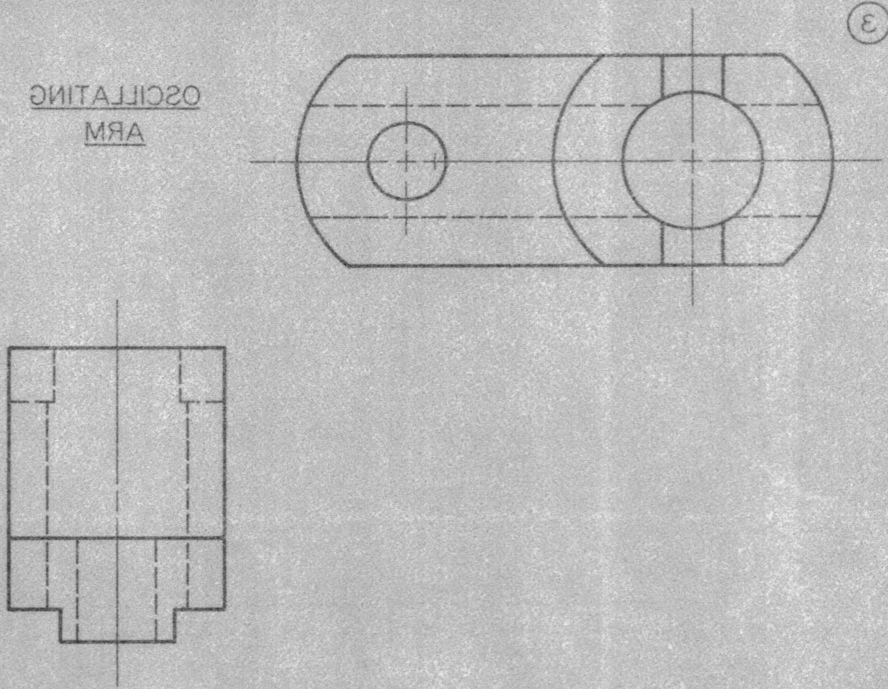

(4)

INDICATER ROLLER

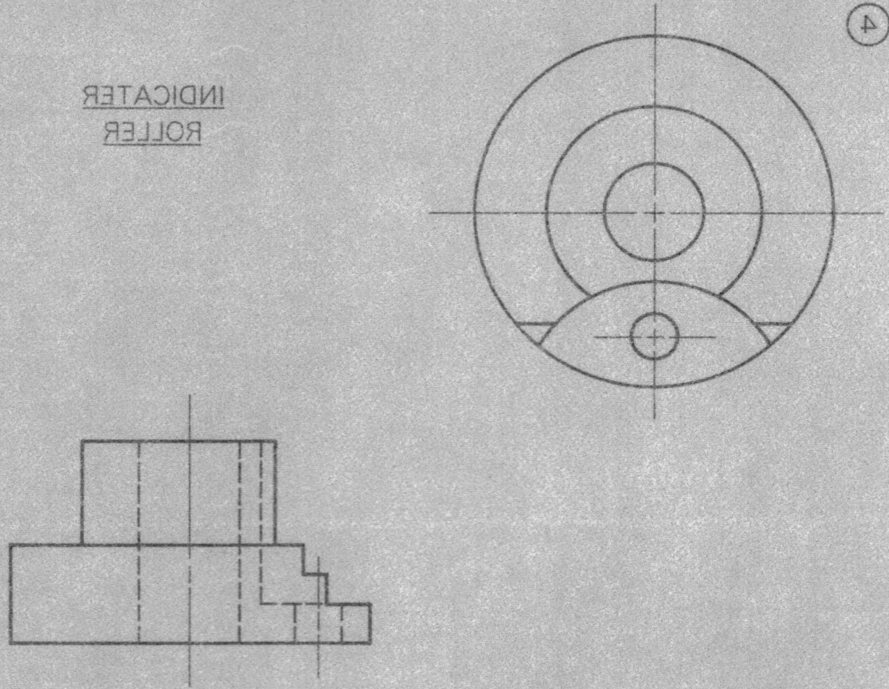

SUPPORT BRACKET

Draw top view.

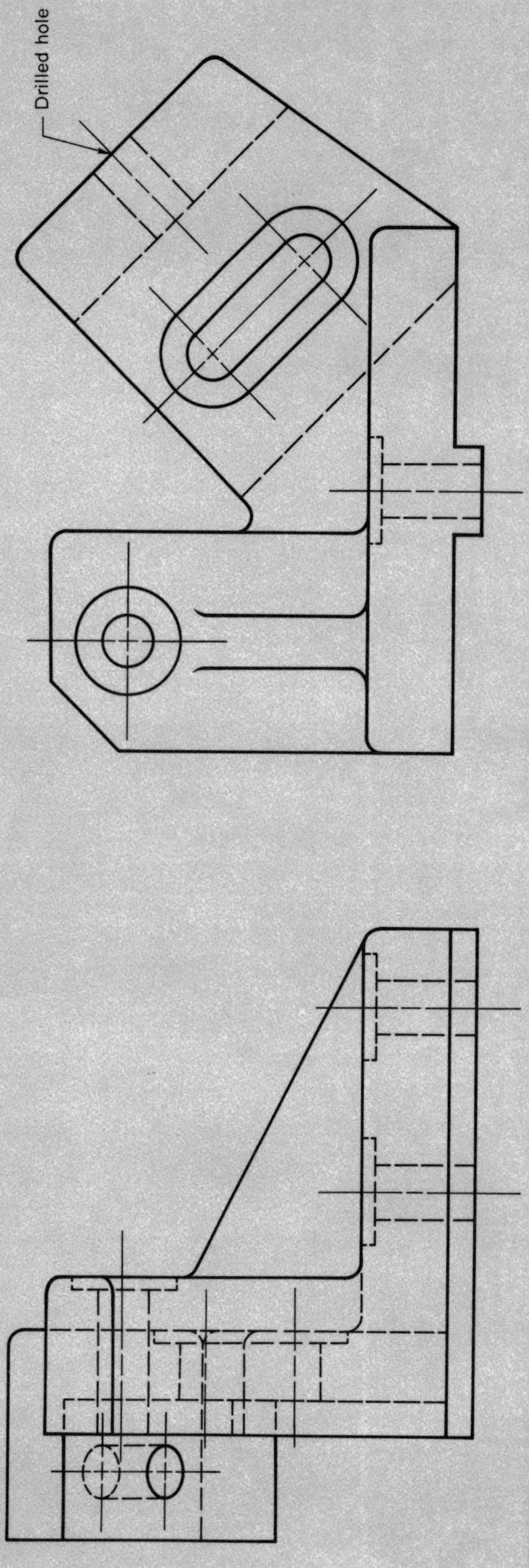

Drilled hole

BOTTOM OF DRAWING

MISSING VIEW MULTIVIEW PROJECTION	DRAWN BY	FILE NO.	DRAWING 22

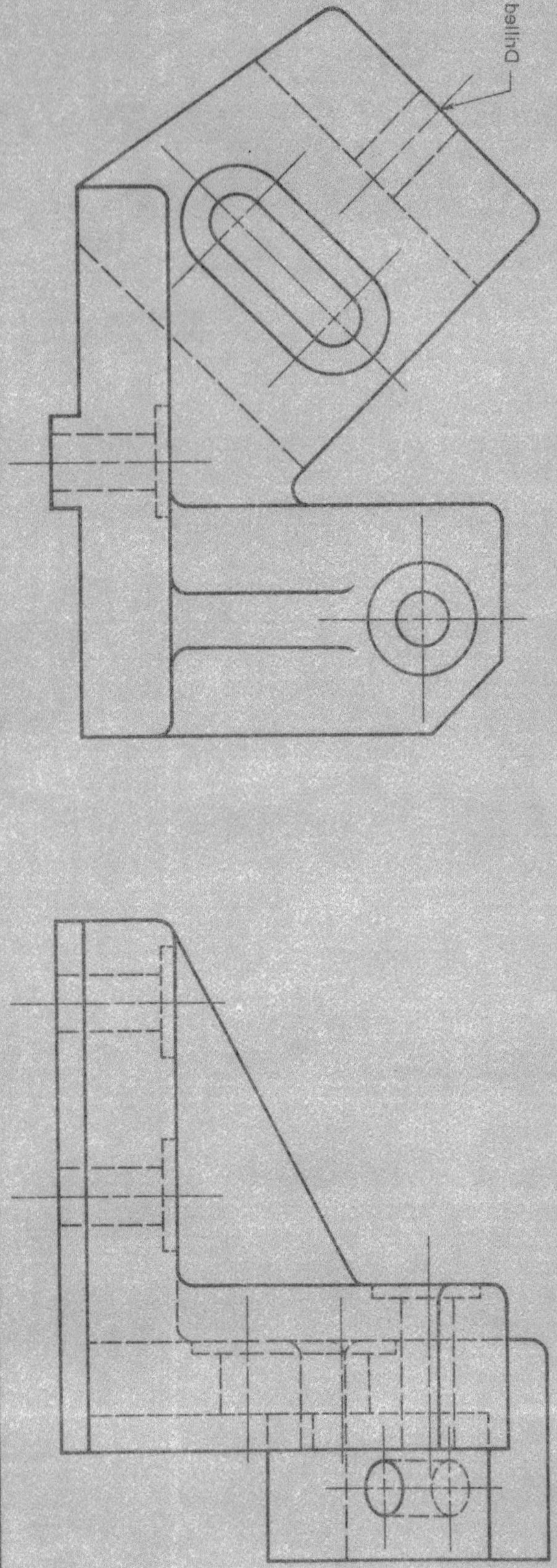
Drilled hole
BOTTOM OF DRAWING

SUPPORT BRACKET

Draw top view.

MULTIVIEW PROJECTION	DRAWN BY	FILE NO.	DRAWING
MISSING VIEW			22

(1)

DRIVE COVER

Draw full section.

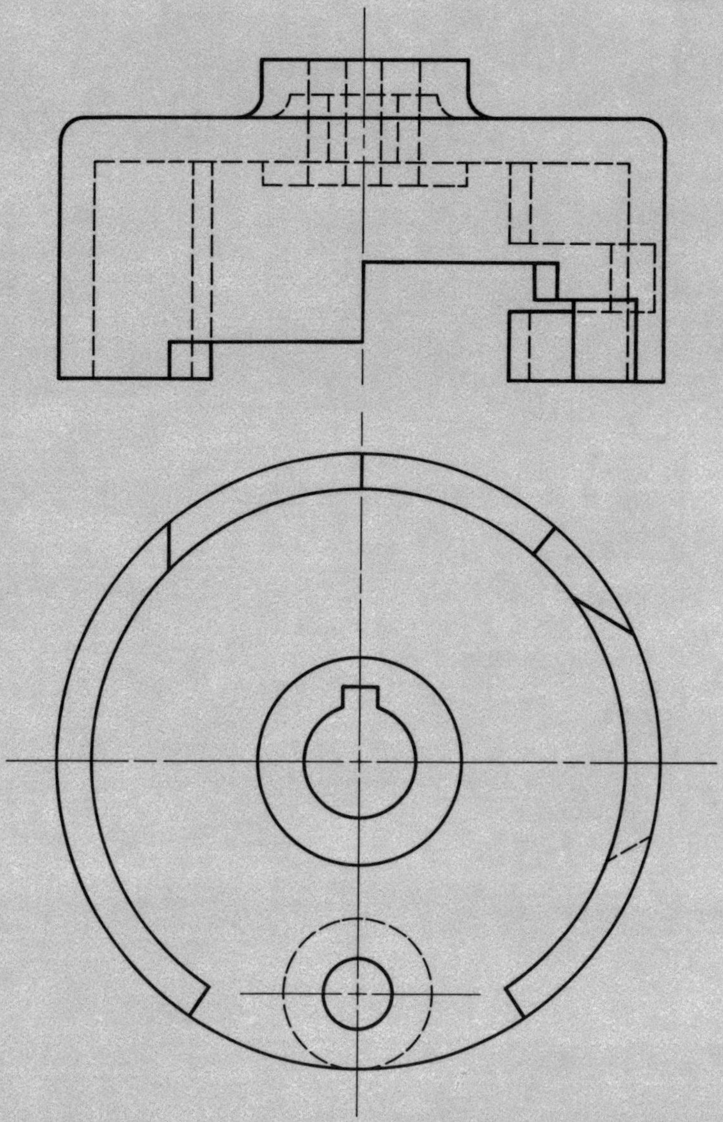

(2)

ASSEMBLY

Complete the full section, using symbolic section lining.

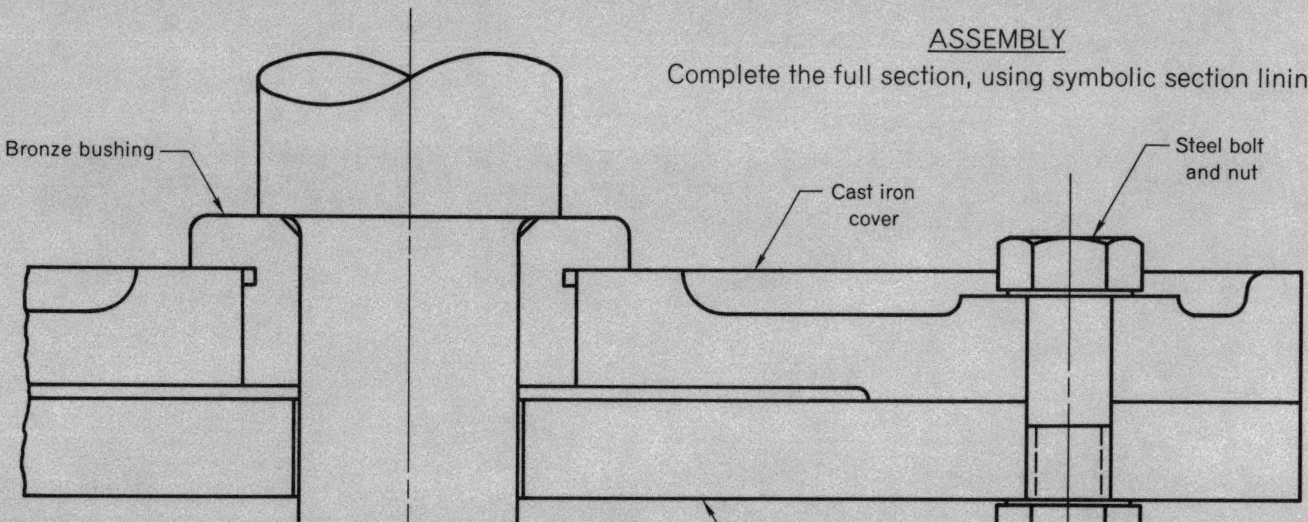
Bronze bushing
Cast iron cover
Steel bolt and nut
Steel plate
Steel shaft

FULL AND ASSEMBLY SECTIONAL VIEWS	DRAWN BY	FILE NO.	DRAWING 26

(1)

DRIVE COVER

Draw full section.

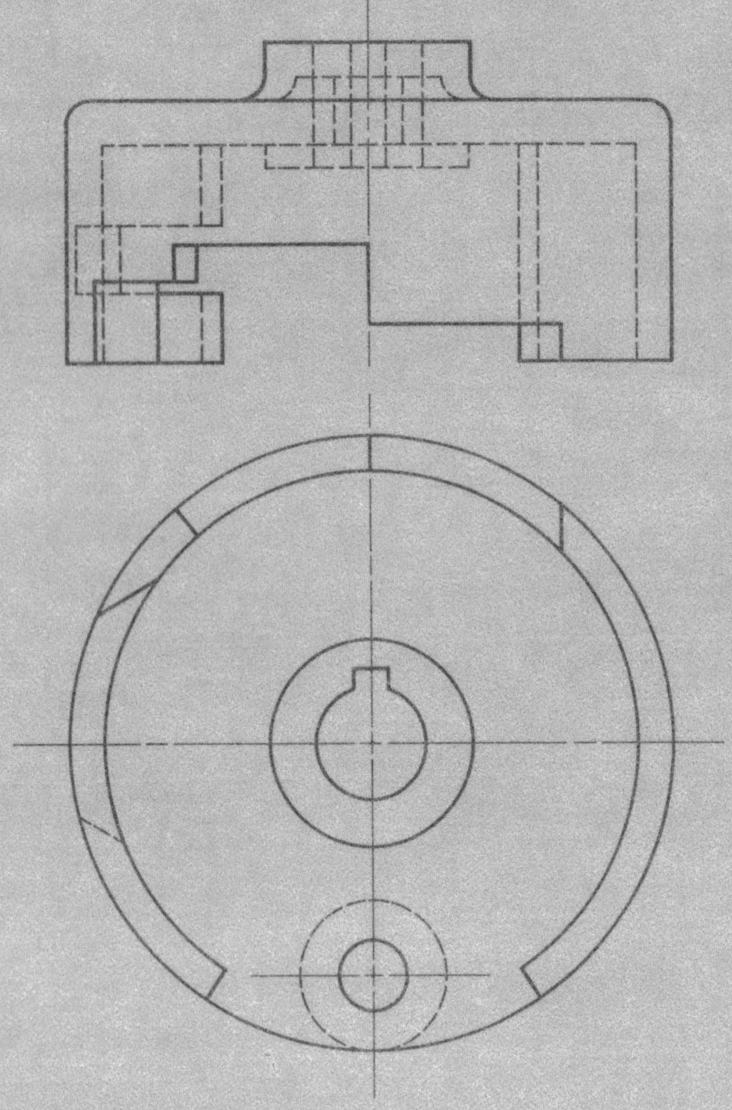

(2)

ASSEMBLY

Complete the full section, using symbolic section lining.

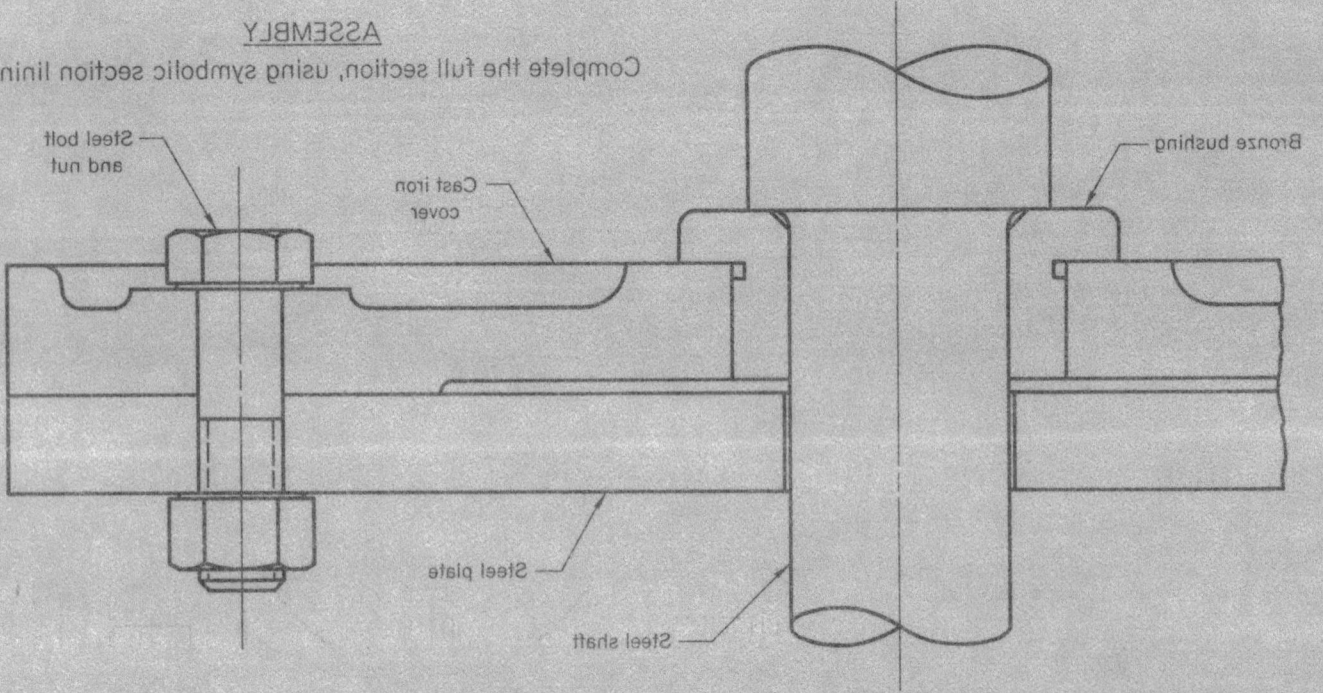

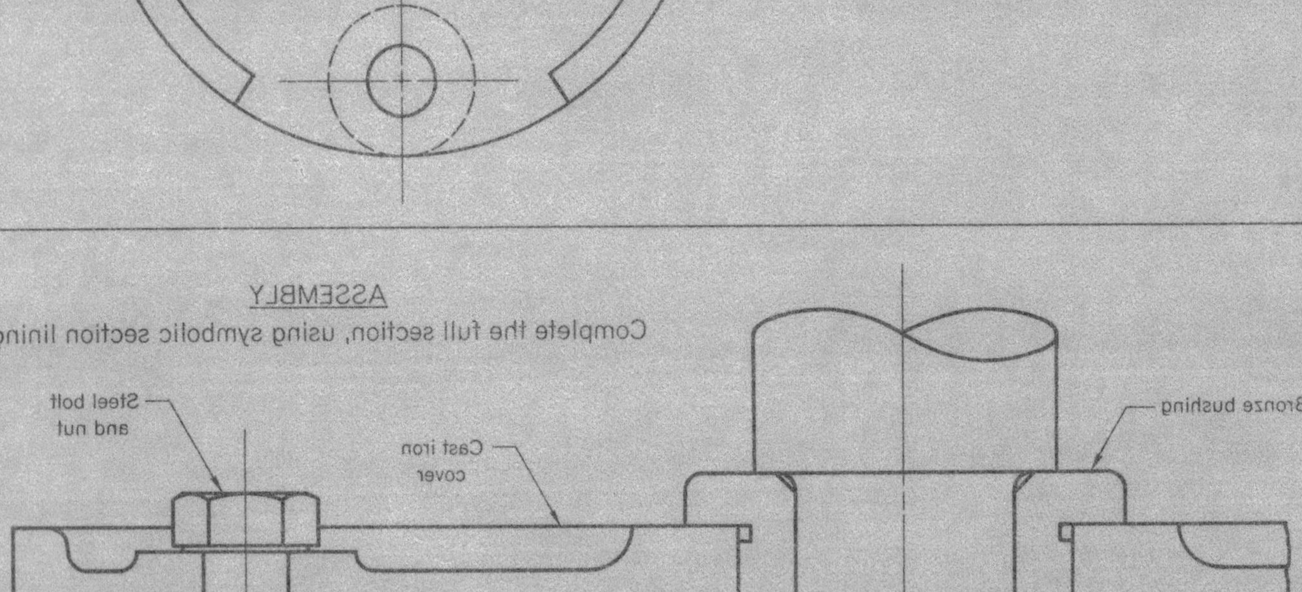
Steel bolt and nut
Cast iron cover
Bronze bushing
Steel plate
Steel shaft

DRAWING	FILE NO.	DRAWN BY	FULL AND ASSEMBLY
26			SECTIONAL VIEWS

BEVEL HOUSING

Draw aligned sections A-A and B-B.

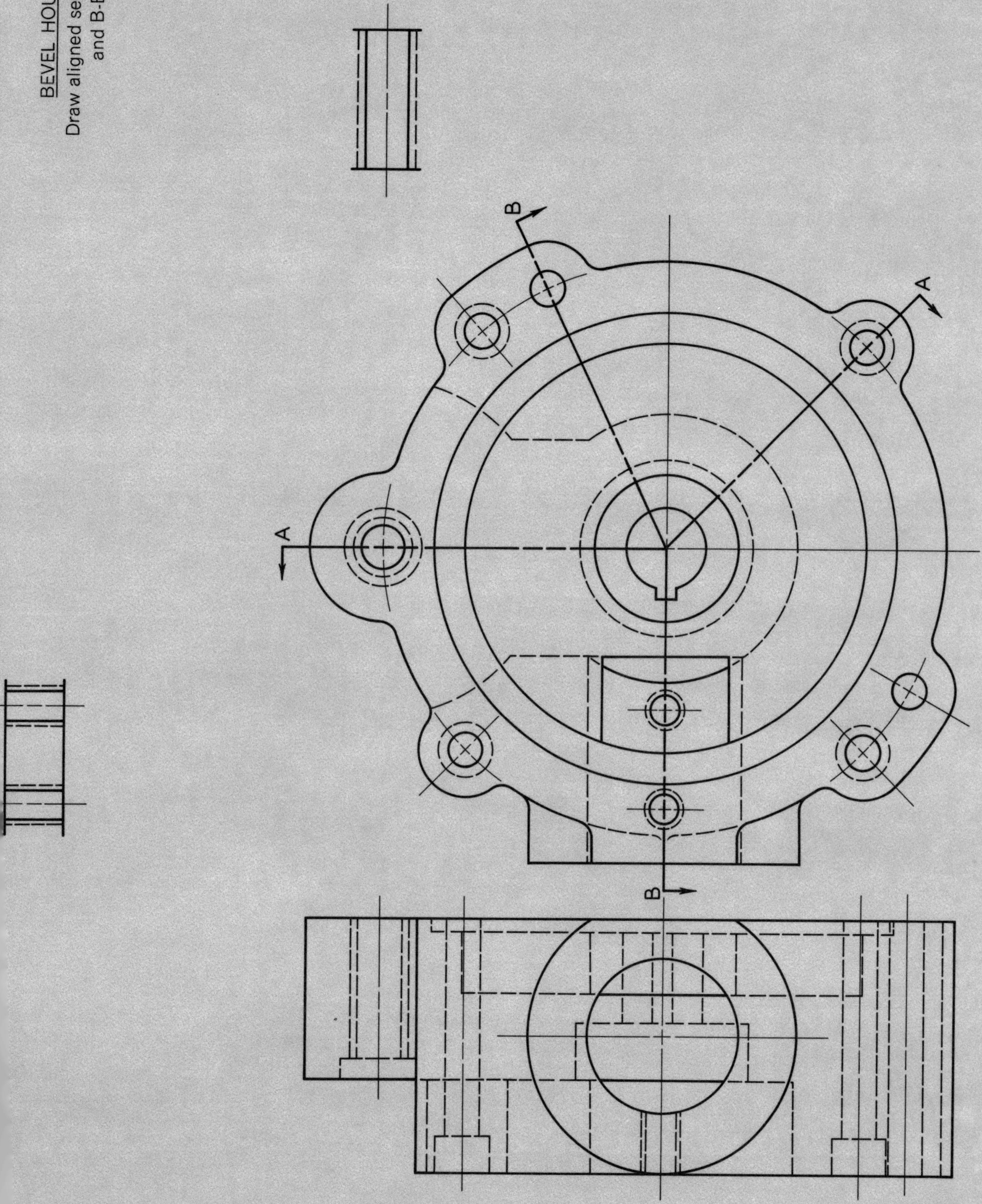

A
A
B
B
BOTTOM OF DRAWING

ALIGNED

SECTIONAL VIEWS

DRAWN BY	FILE NO.	DRAWING
		27

BEVEL HOUSING

Draw aligned sections A-A and B-B.

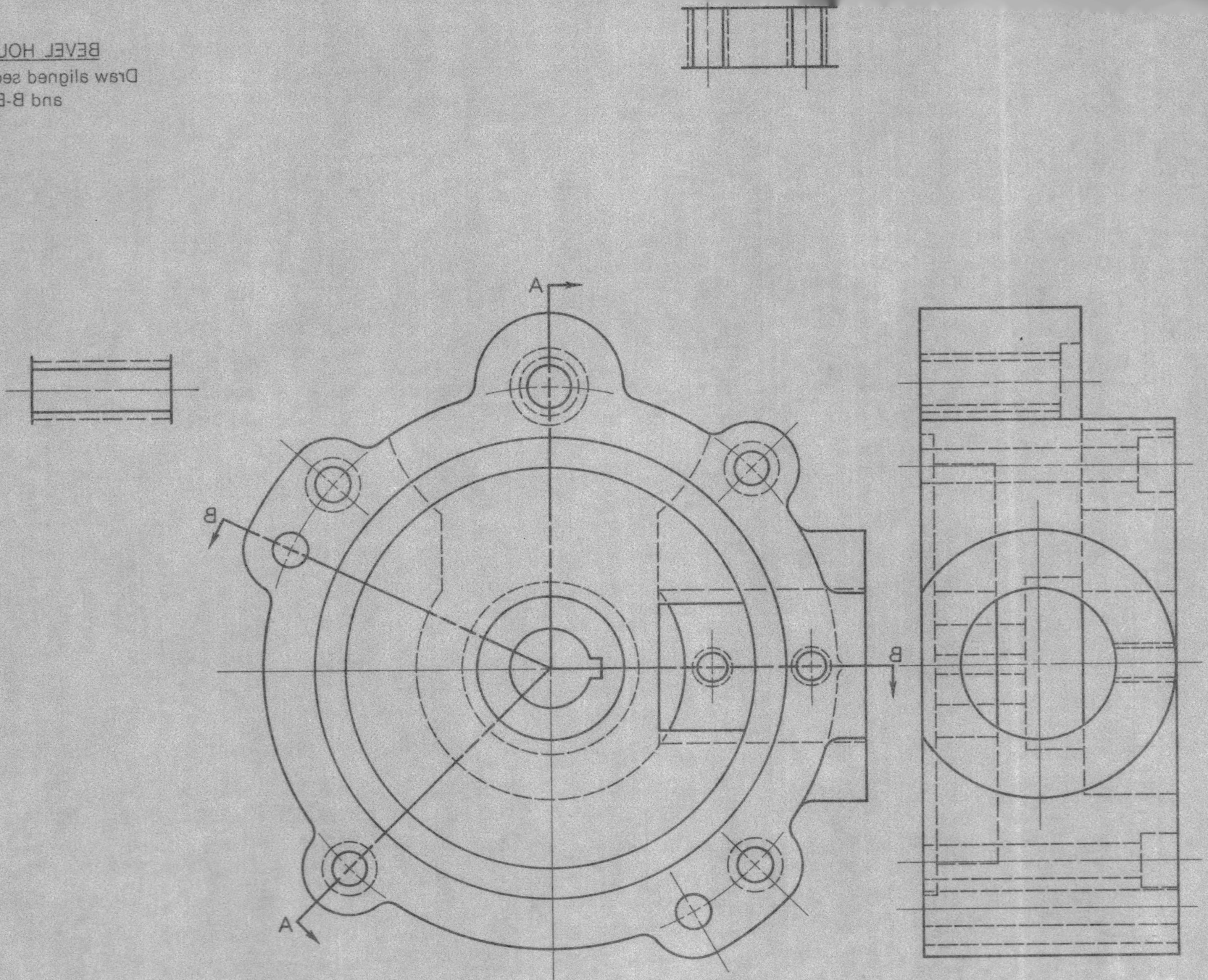

A
A
B
B

BOTTOM OF DRAWING

ALIGNED SECTIONAL VIEWS	DRAWN BY	FILE NO.	DRAWING 27

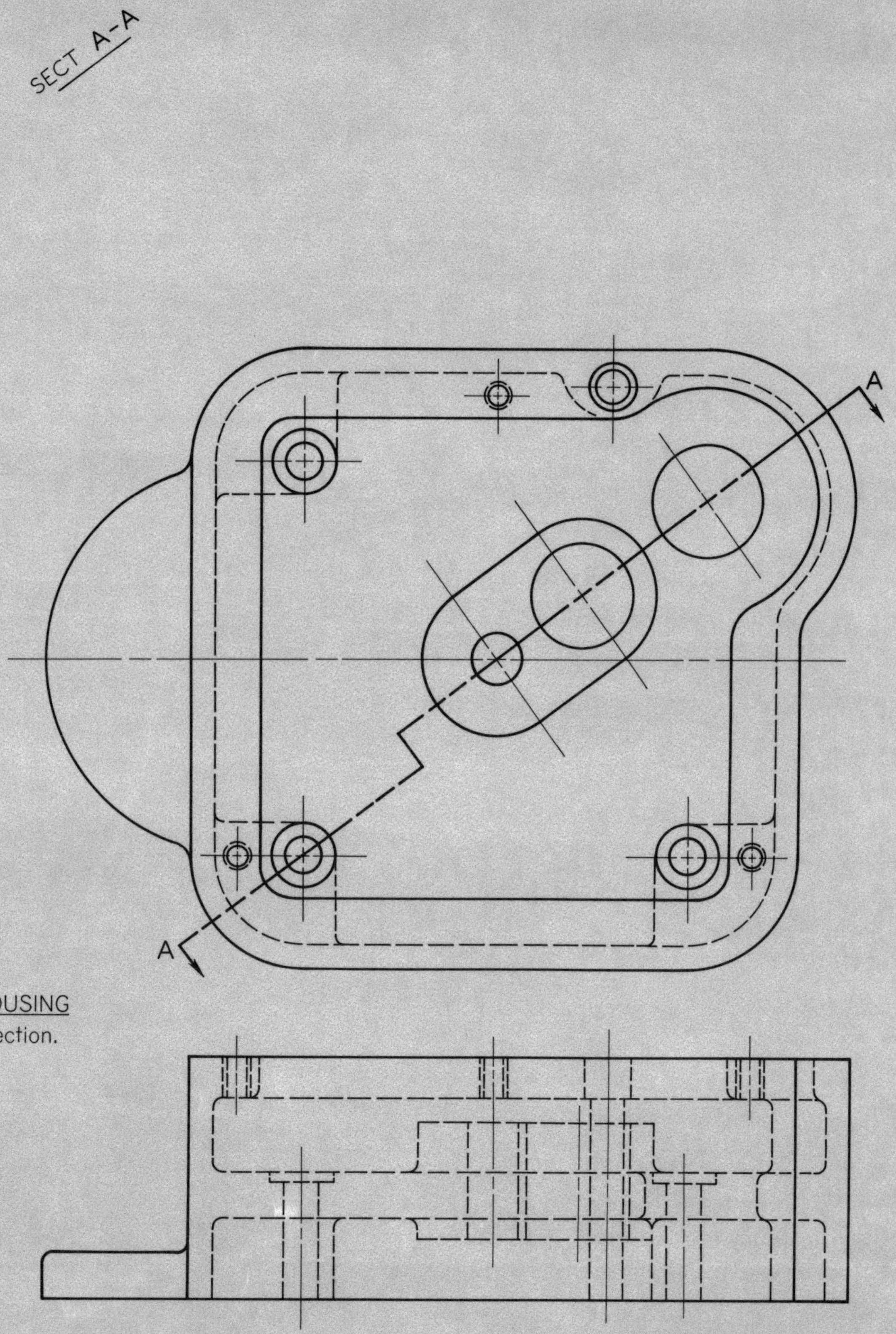
SECT A-A
A
A

HEADSTOCK HOUSING

Draw auxiliary section.

OFFSET SECTION AUXILIARY VIEW	DRAWN BY	FILE NO.	DRAWING 31

SECT A-A

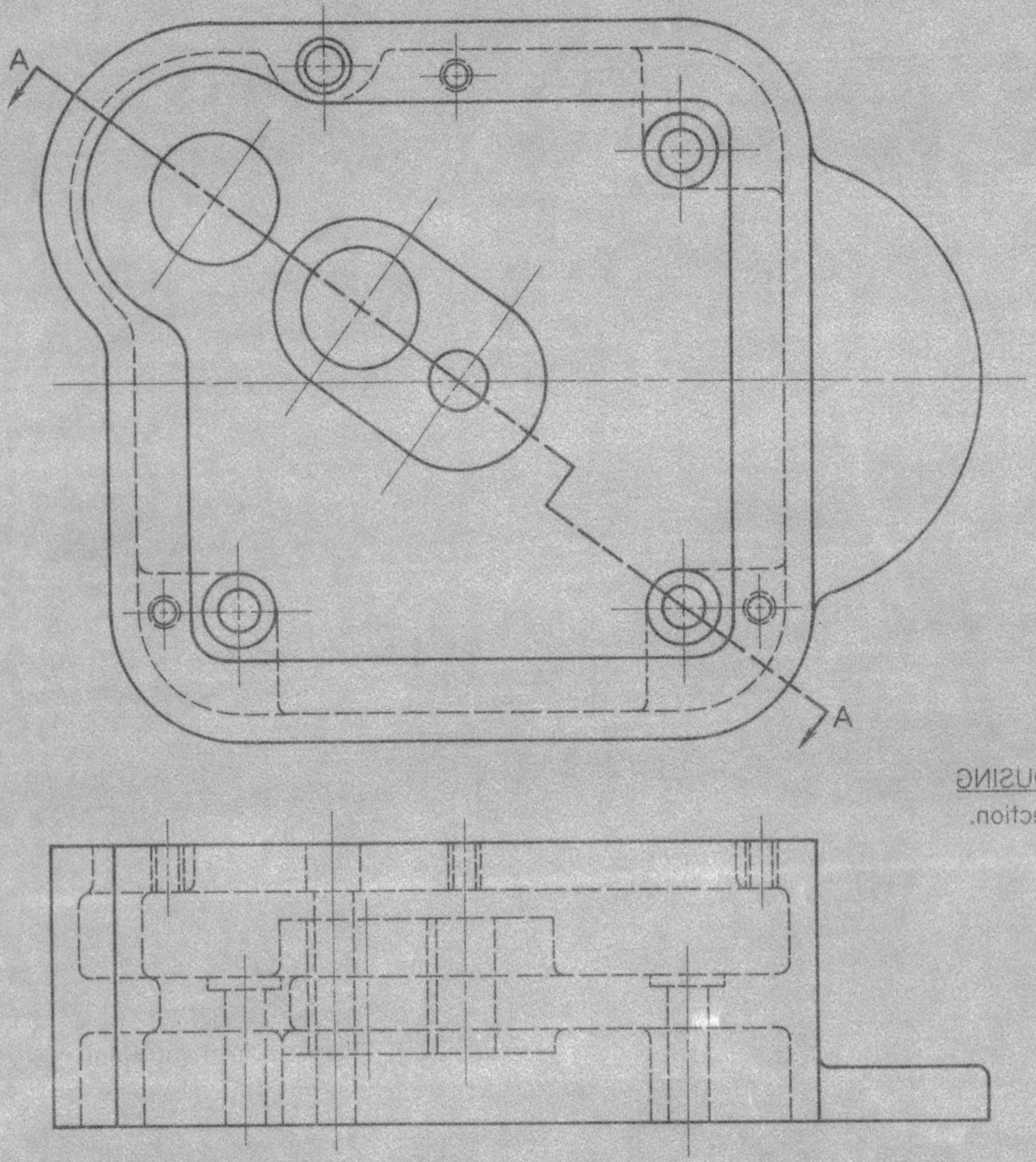
A
A

HEADSTOCK HOUSING

Draw auxiliary section.

DRAWING	FILE NO.	DRAWN BY	OFFSET SECTION
31			AUXILIARY VIEW

①

LOCK CATCH

Draw isometric drawing.

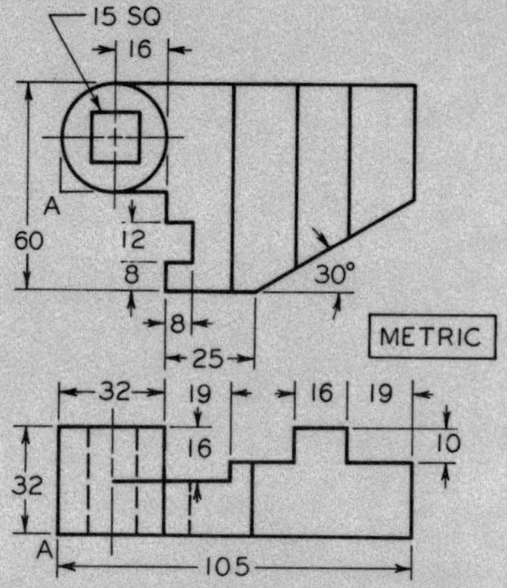

15 SQ
16
A
60
12
8
8
25
30°
METRIC
32
19
16
19
16
10
32
A
105

A

②

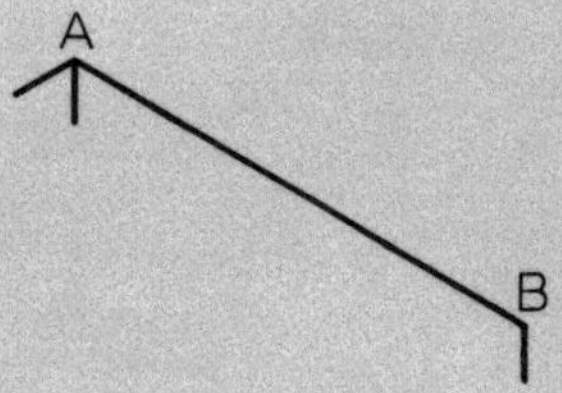

A
B

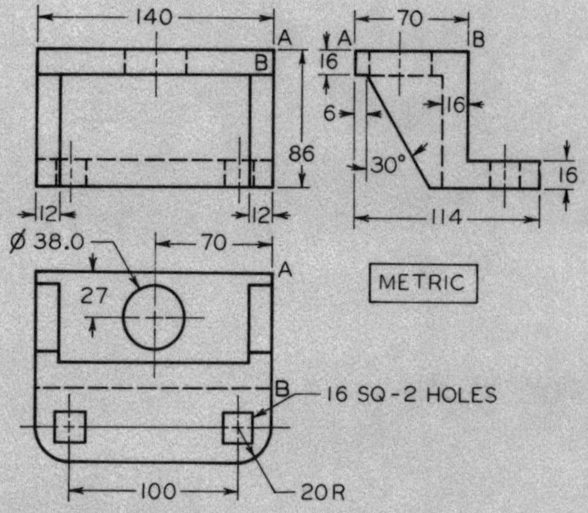

140
A
B
86
12
12
70
A
B
16
6
16
30°
16
114
Ø 38.0
70
A
27
METRIC
B
16 SQ-2 HOLES
100
20R

BRACKET

Draw isometric drawing.
Half size.

MECHANICAL ISOMETRIC DRAWING	DRAWN BY	FILE NO.	DRAWING 37

1

LOCK CATCH

Draw isometric drawing.

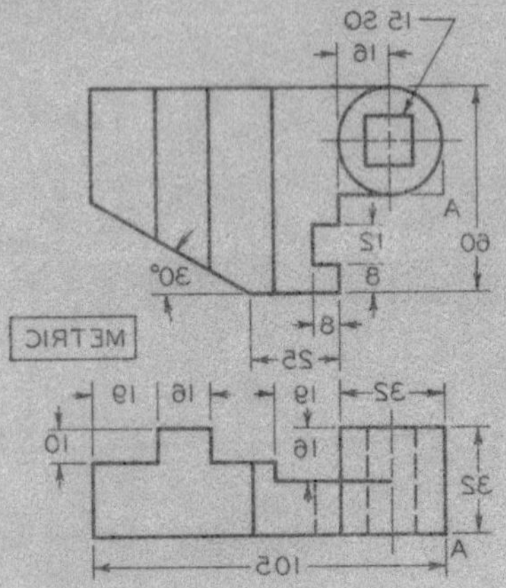

15 SQ
16
A
60
12
8
30°
METRIC
8
25
32
19
16
19
10
16
32
A
105

A

2

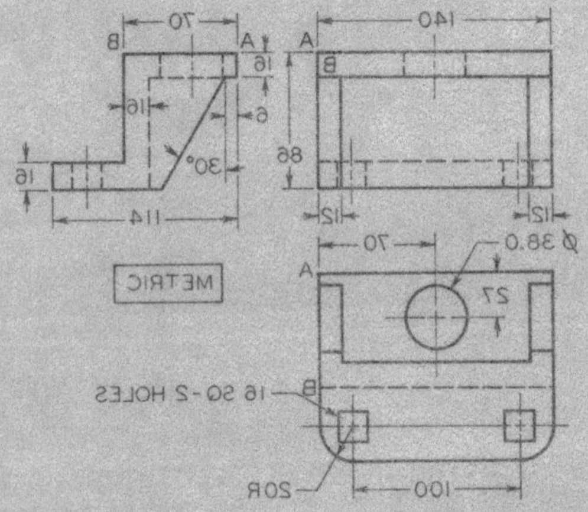

70
B
A
16
16
6
30°
16
114
METRIC
140
A
B
86
12
12
70
Ø 38.0
A
27
16 SQ - 2 HOLES
B
20R
100

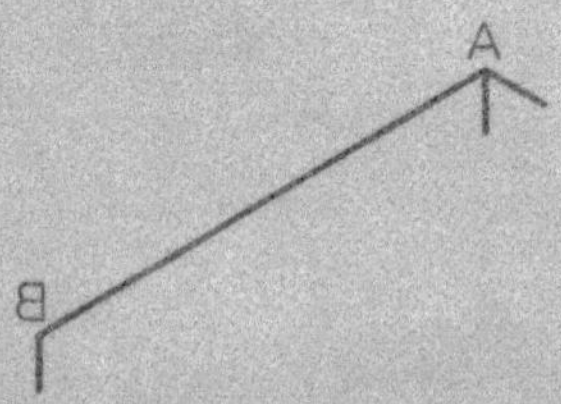

A
B

BRACKET

Draw isometric drawing.
Half size.

MECHANICAL ISOMETRIC DRAWING	DRAWN BY	FILE NO.	DRAWING 37

This sheet and the following sheet contain the views of the parts, a standard parts list, and an assembly pictorial of the Roller Guide. Add dimensions, notes, etc., to complete the details. Use RC 6 limits for large hole in base.

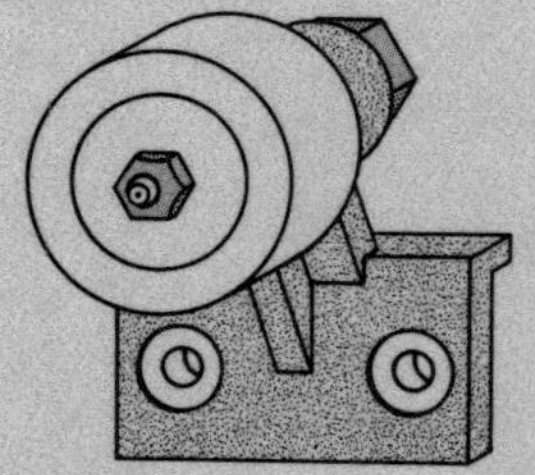

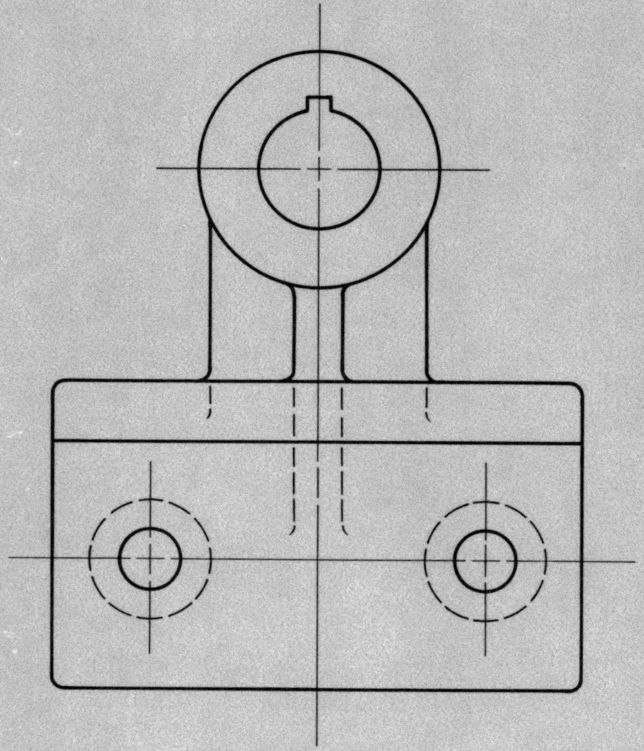

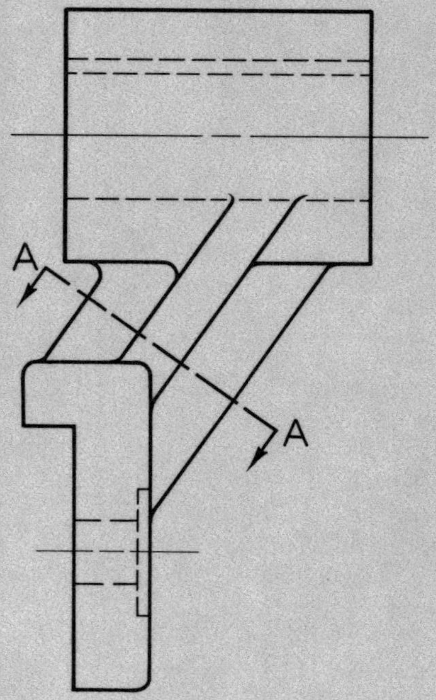

A
A

(1) BASE

FOR
ROLLER GUIDE

CAST IRON
1 REQD

SCALE: 1 = 1

METRIC

FILLETS AND ROUNDS 1.5 R

SECTION A-A

BOTTOM OF DRAWING

MATING PARTS	DRAWN BY	FILE NO.	DRAWING
DIMENSIONING			42

This sheet and the following sheet contain the views of the parts, a standard parts list, and an assembly pictorial of the Roller Guide. Add dimensions, notes, etc., to complete the details. Use RC 6 limits for large hole in base.

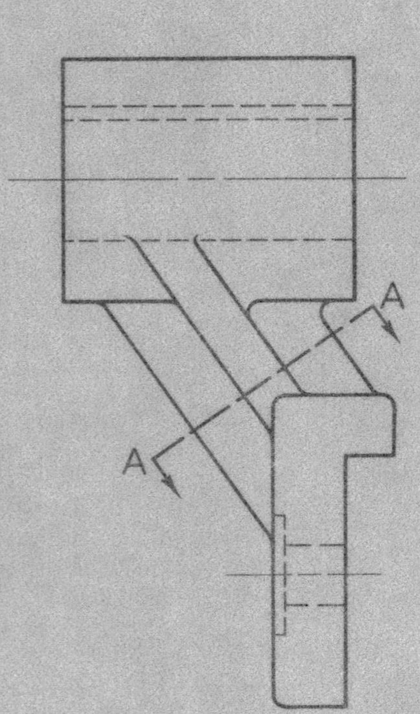
A
A

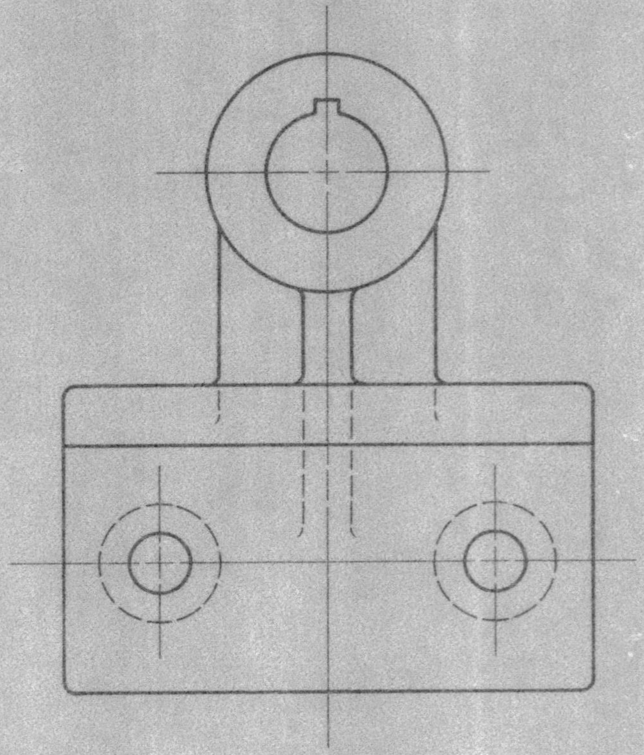

(1) BASE
FOR
ROLLER GUIDE
CAST IRON
1 REQD
SCALE: 1 = 1

SECTION A-A

FILLETS AND ROUNDS 1.5 R

METRIC

DIMENSIONING MATING PARTS	DRAWN BY	FILE NO.	DRAWING 42

BOTTOM OF DRAWING

STANDARD PARTS

(5) 1 - M16 × 2 HEXAGON NUT

(6) 1 - $\frac{11}{16}$ - AMERICAN NATIONAL STD REGULAR LOCKWASHER

(7) 1 - NO. 8585 HYDRAULIC GREASE FITTING

(8) 1 - NO. 405 WOODRUFF KEY

SCHEDULE OF FITS FOR BUSHING

Inside diameter	RC 3
Outside diameter	FN 2

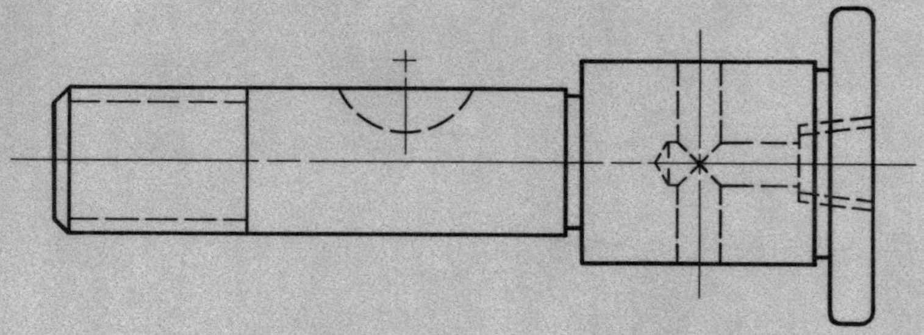

(2) SPECIAL BOLT CRS - 1 REQD

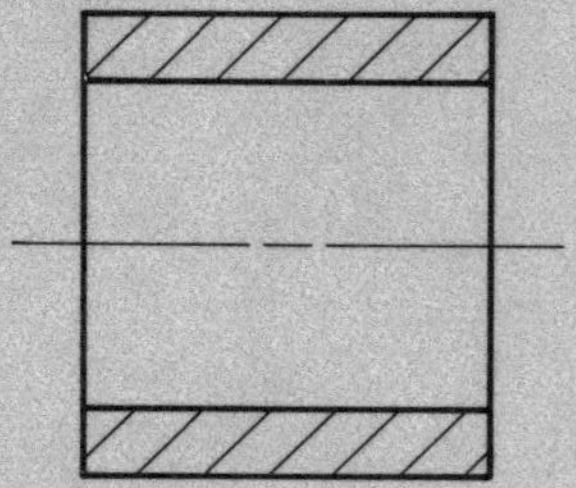

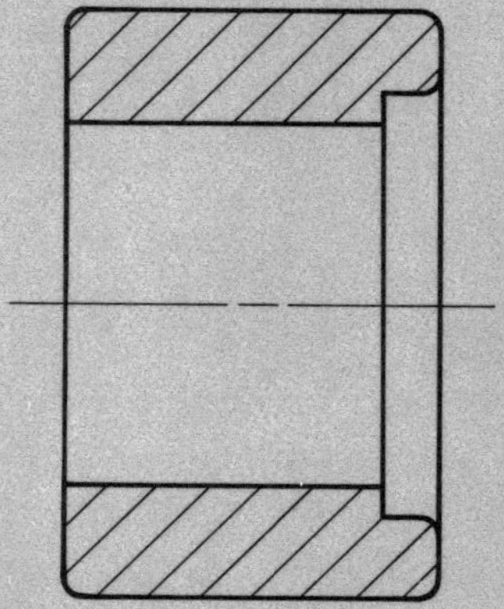

(4) BUSHING
BRZ - 1 REQD

(3) ROLLER
CRS - 1 REQD

FILLETS AND ROUNDS 1.5 R
SCALE: 1 = 1

METRIC

BOTTOM OF DRAWING

MATING PARTS DIMENSIONING	DRAWN BY	FILE NO.	DRAWING 43

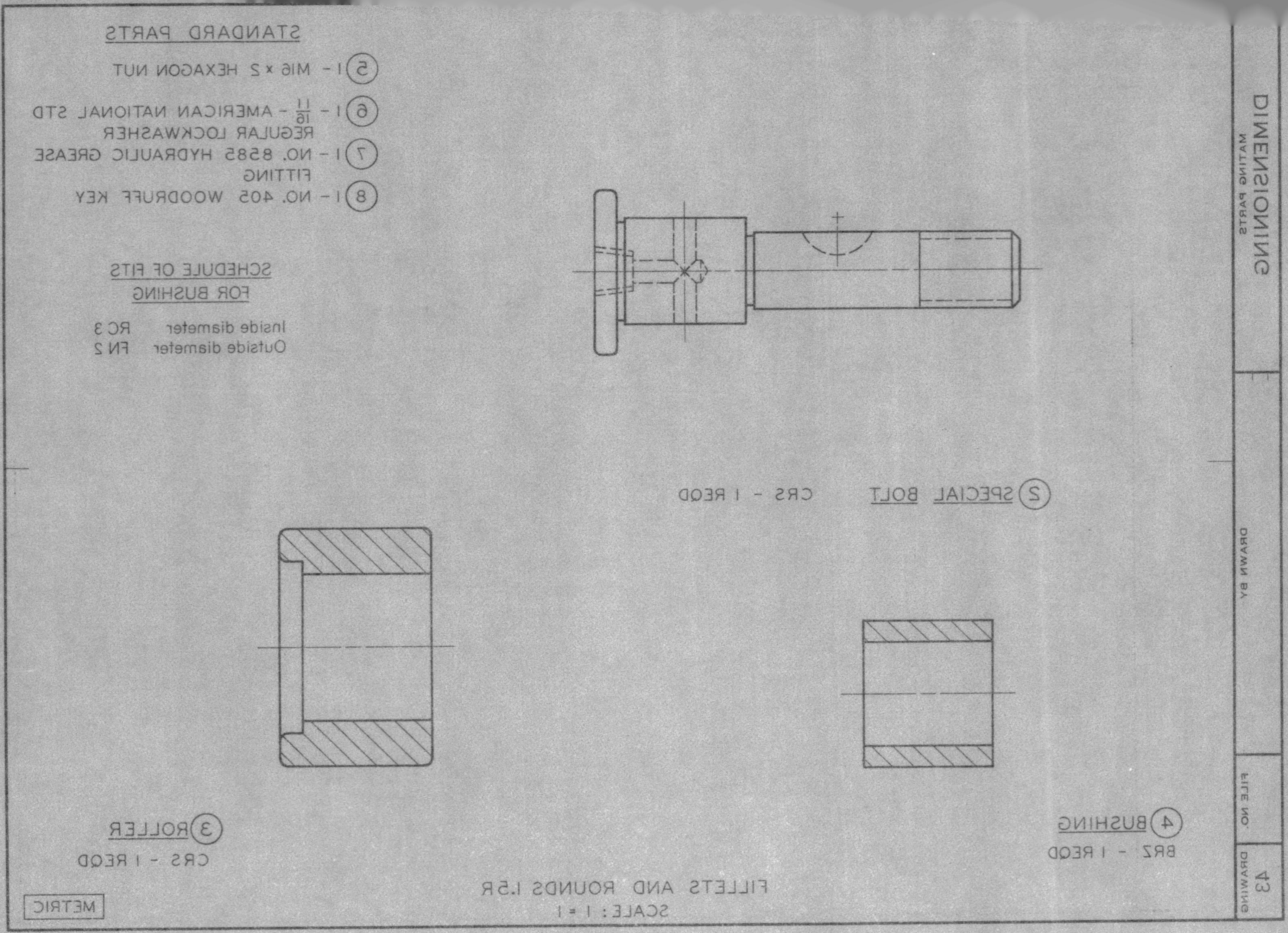
STANDARD PARTS
5 1 - M16 × 2 HEXAGON NUT
6 1 - 11/16 - AMERICAN NATIONAL STD REGULAR LOCKWASHER
7 1 - NO. 8585 HYDRAULIC GREASE FITTING
8 1 - NO. 405 WOODRUFF KEY
SCHEDULE OF FITS FOR BUSHING
Inside diameter RC 3
Outside diameter FN 2
2 SPECIAL BOLT CRS - 1 REQD
3 ROLLER CRS - 1 REQD
4 BUSHING BRZ - 1 REQD
FILLETS AND ROUNDS 1.5R
SCALE: 1 = 1
METRIC
BOTTOM OF DRAWING
MATING PARTS
DIMENSIONING
DRAWN BY
FILE NO.
DRAWING
43

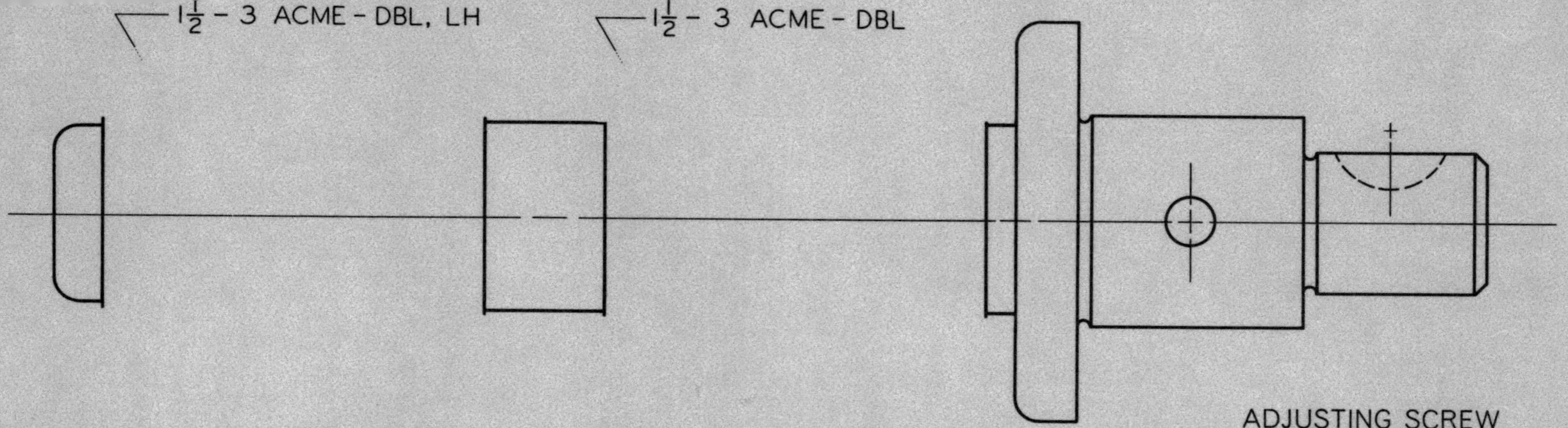

$1\frac{1}{2}$ - 3 ACME - DBL, LH
$1\frac{1}{2}$ - 3 ACME - DBL

ADJUSTING SCREW

FULL SIZE

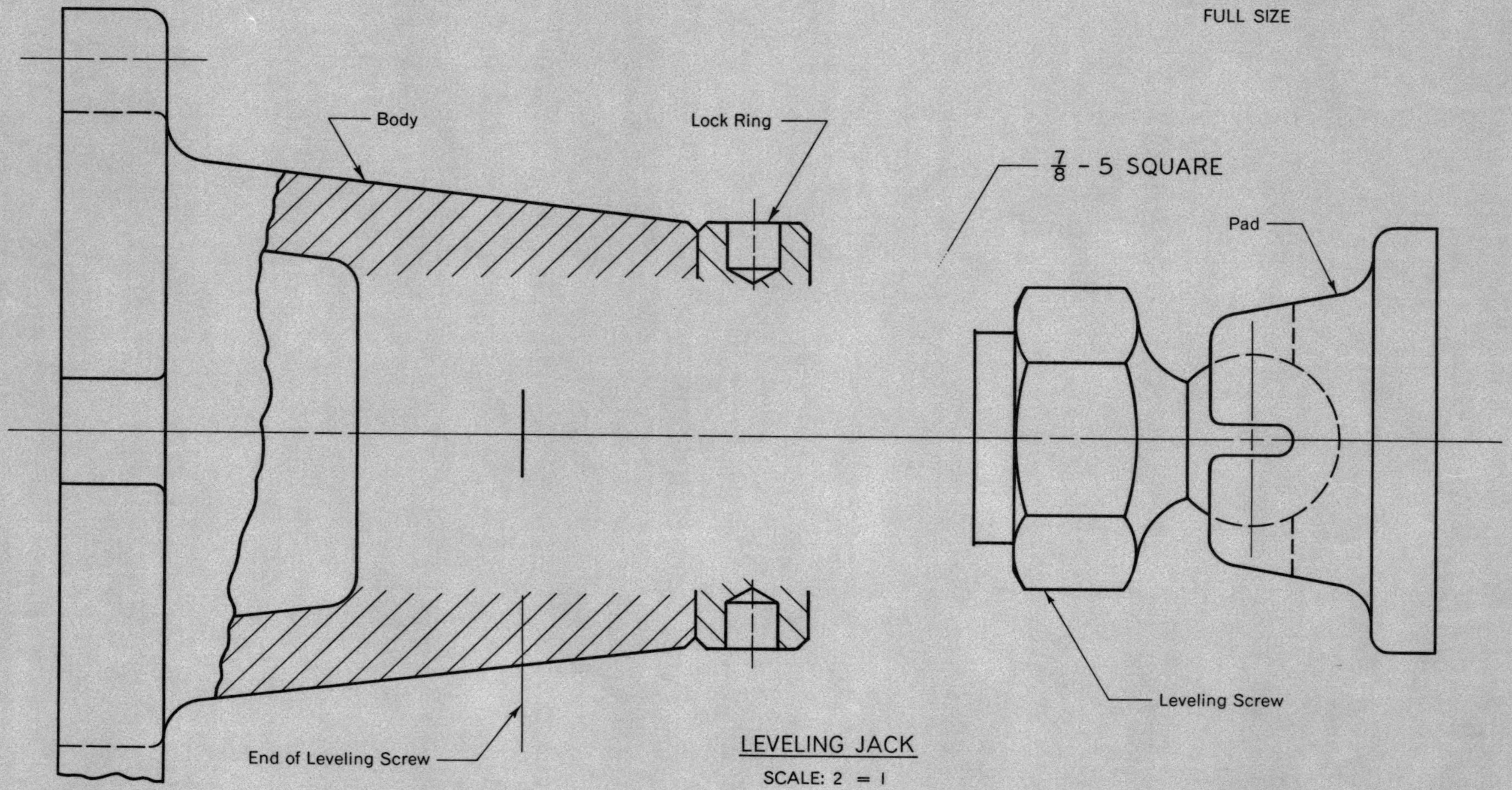

Body
Lock Ring
$\frac{7}{8}$ - 5 SQUARE
Pad
Leveling Screw
End of Leveling Screw

LEVELING JACK

SCALE: 2 = 1

Complete the views as specified.

DETAILED	DRAWN BY	FILE NO.	DRAWING
CME & SQUARE THREADS			46

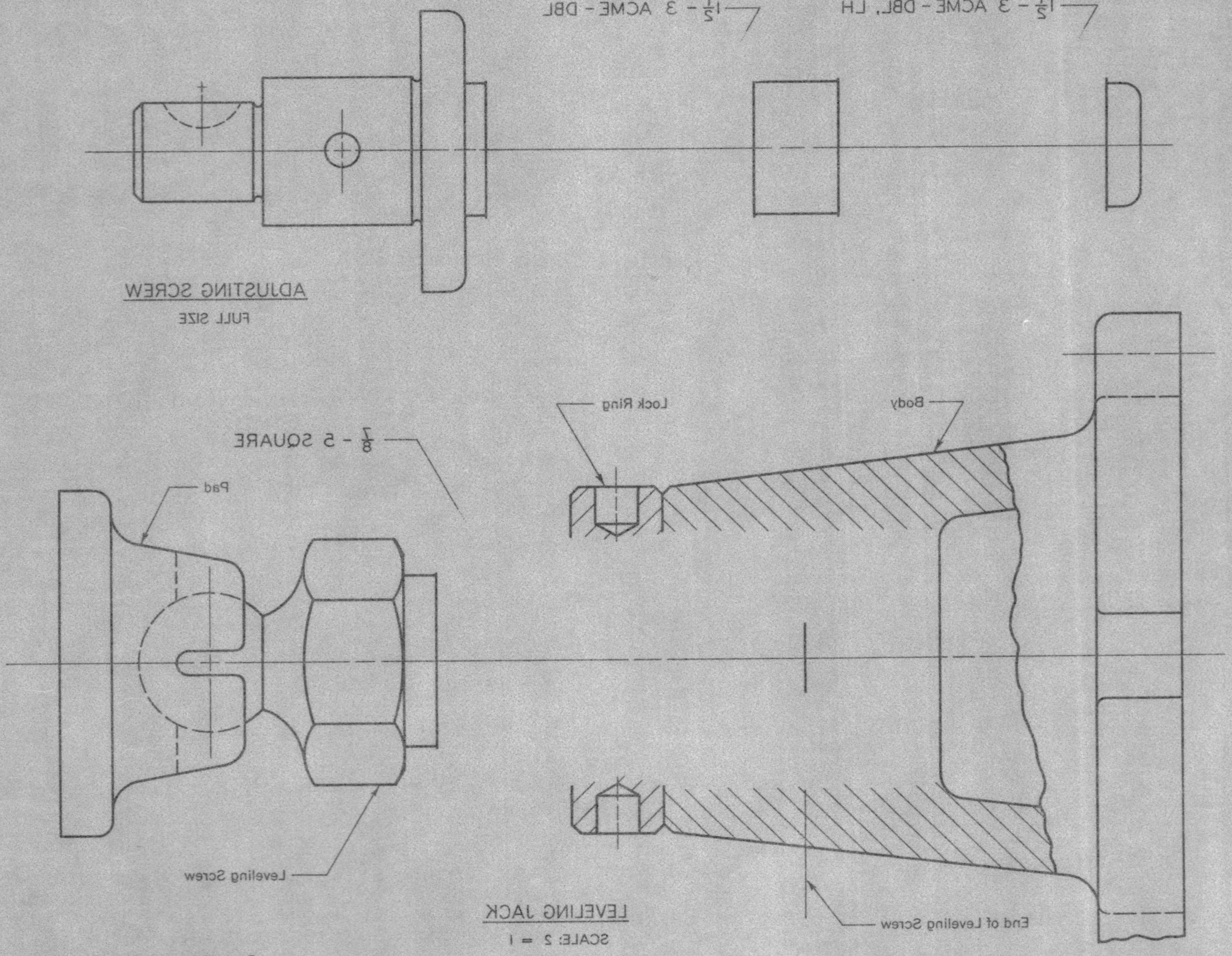
1½ - 3 ACME - DBL
1½ - 3 ACME - DBL, LH
ADJUSTING SCREW
FULL SIZE
⅞ - 5 SQUARE
Pad
Lock Ring
Body
Leveling Screw
End of Leveling Screw
LEVELING JACK
SCALE: 2 = 1

Complete the views as specified.

DETAILED
ACME & SQUARE THREADS
DRAWN BY
FILE NO.
DRAWING
46

Complete the views of the intersecting forms of the Collector.

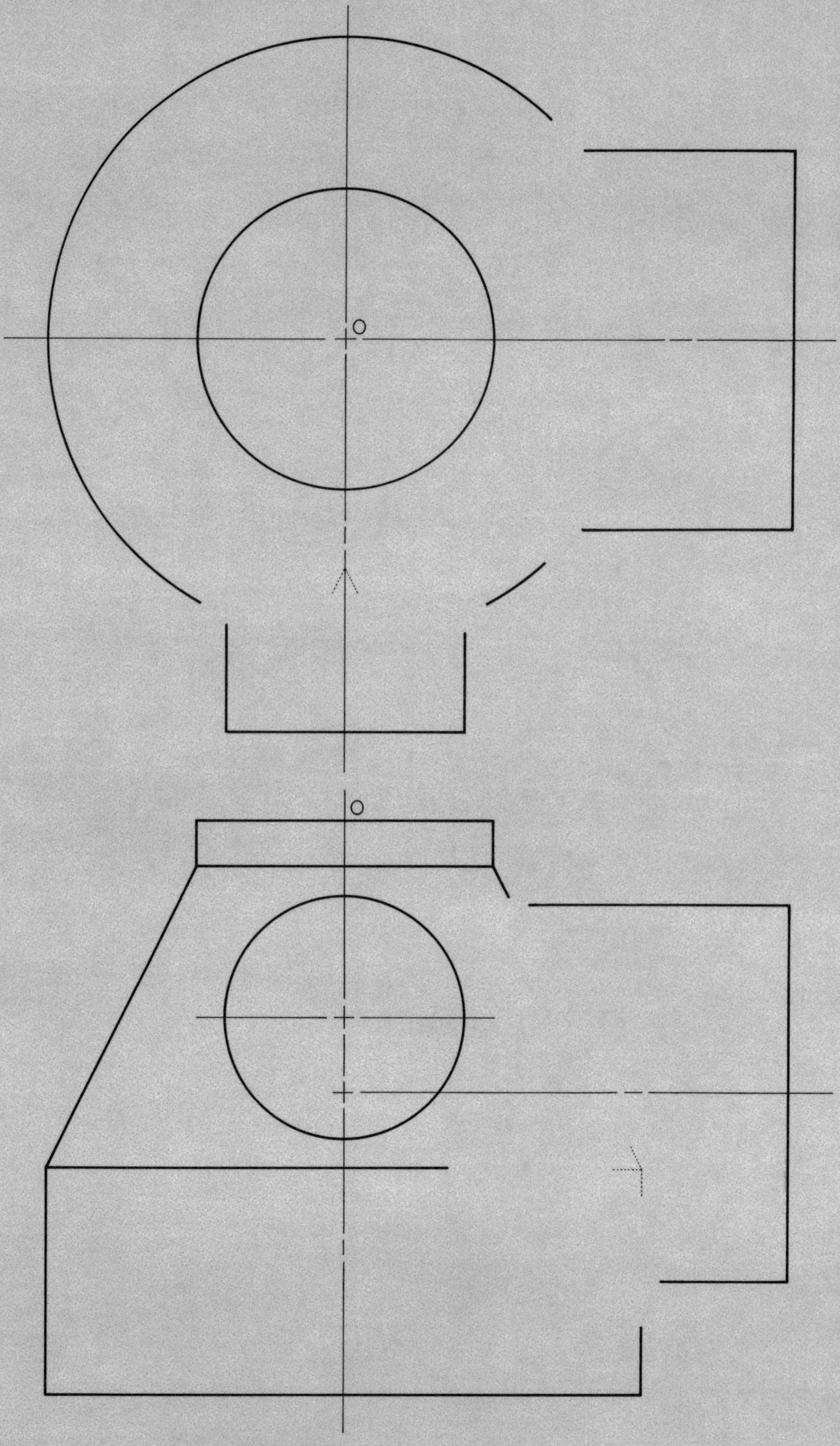

O
O

CIRCULAR FORMS INTERSECTIONS	DRAWN BY	FILE NO.	DRAWING 71

Complete the views of the intersecting forms of the Collector.

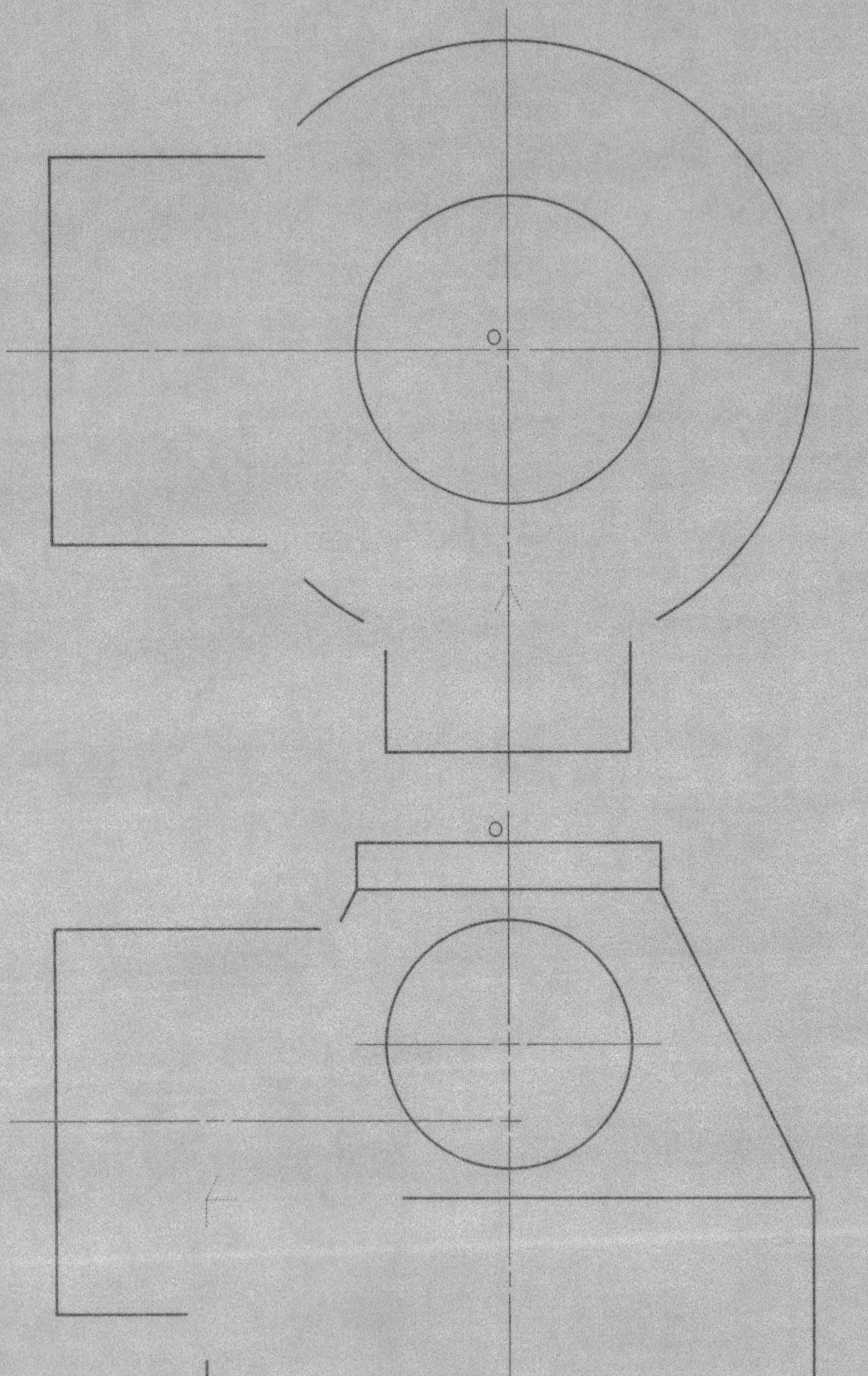

CIRCULAR FORMS INTERSECTIONS	DRAWN BY	FILE NO.	DRAWING 71

... transition piece, starting with seam E-1 at the indicated position and ending at the center line F-7 of panel B-C-7.

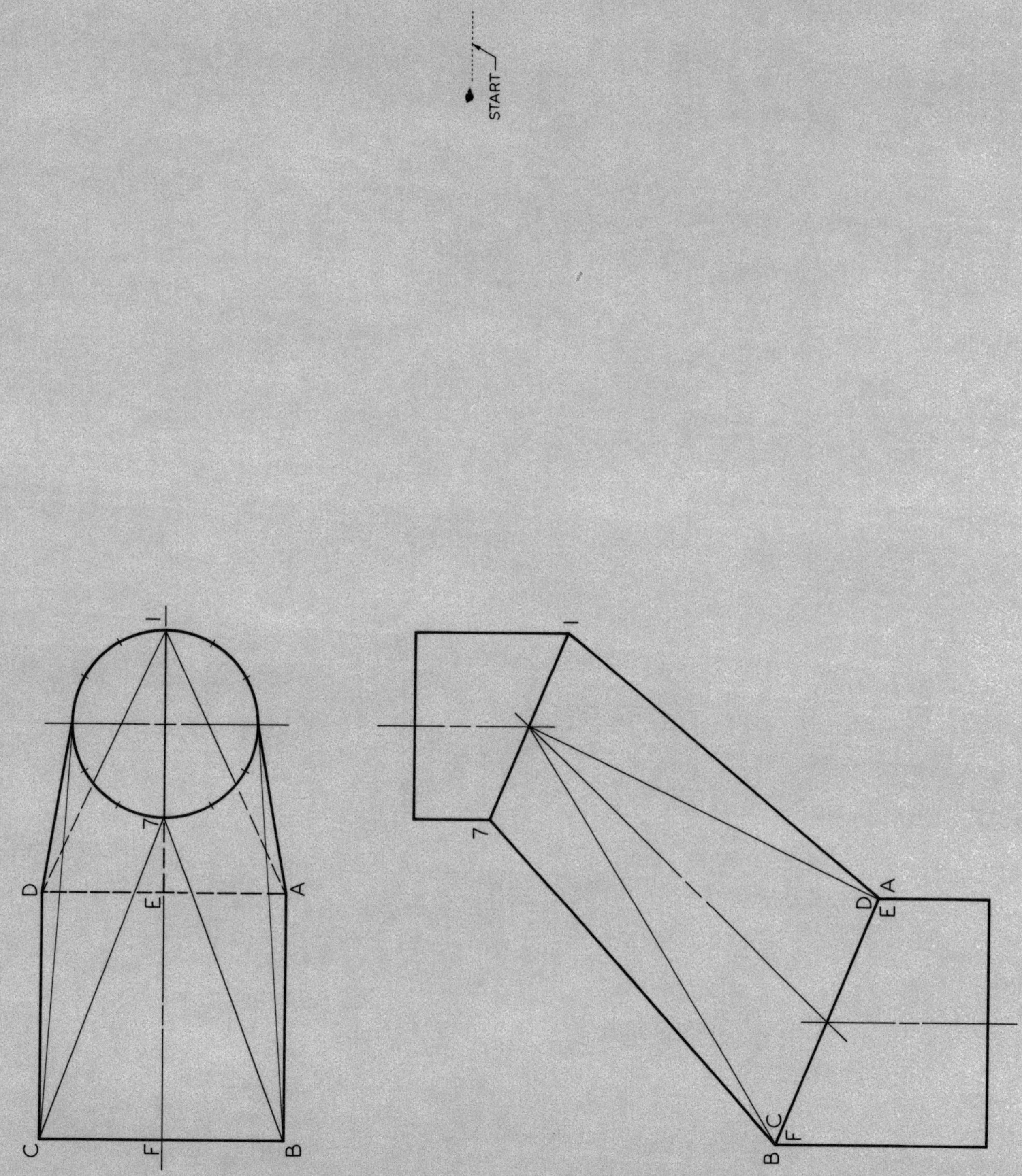

START
I
7
D
E
A
C
F
B

BOTTOM OF DRAWING

TRANSITION PIECE TRIANGULATION	DRAWN BY	FILE NO.	DRAWING 75

... piece, starting with seam E-1
at the indicated position and ending at the center line F-7 of panel B-C-7.

START

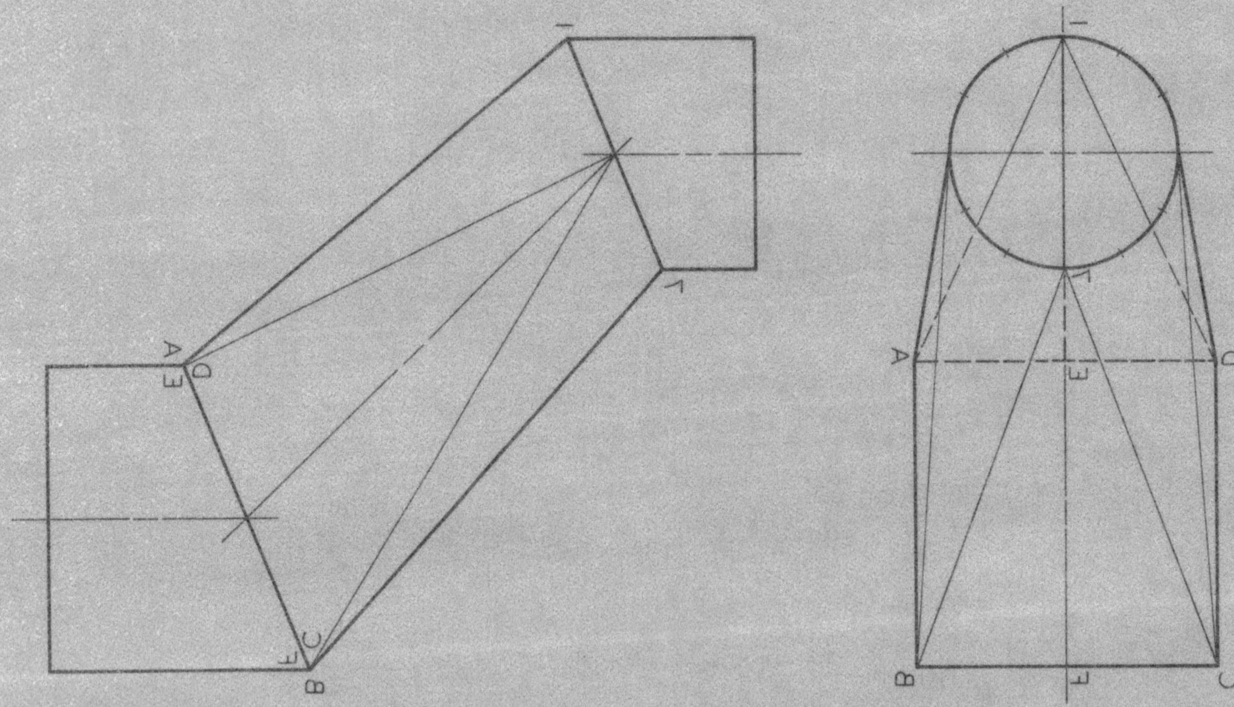

A
B
C
D
E
F
1
7

BOTTOM OF DRAWING

TRANSITION PIECE	DRAWN BY	FILE NO.	DRAWING
TRIANGULATION			75

Plot the cut and fill areas for the level roadway at an elevation of 50 m, with a slope of 1:1 for the cut and 1.5:1 for the fill. Vertical Scale: 1/200. Crosshatch the cut and fill areas in opposing directions. Also crosshatch the rectangles under "KEY" below for identification.

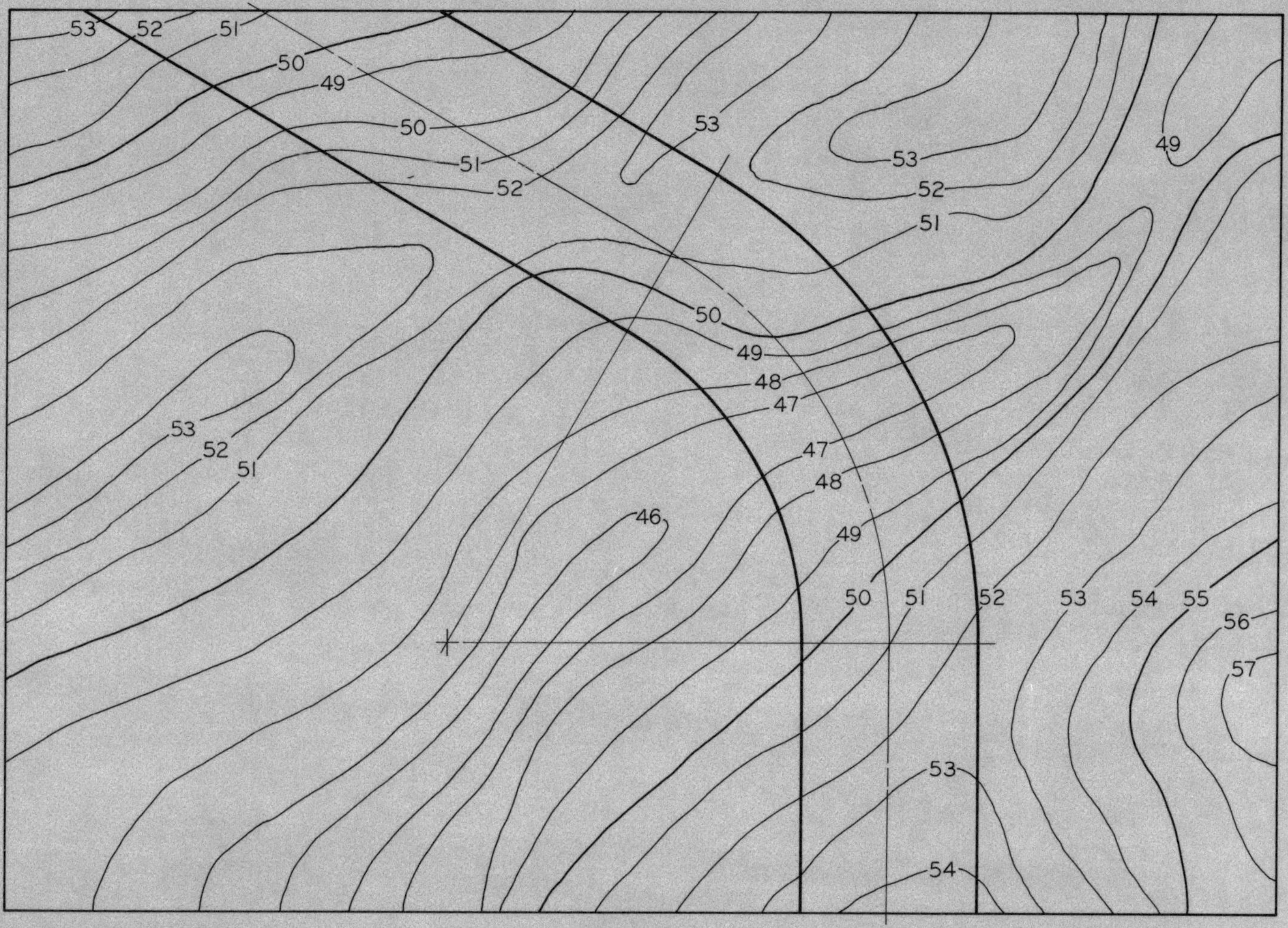

53
52
51
50
49
50
51
52
53
53
52
51
49
50
49
48
47
53
52
51
47
48
46
49
50
51
52
53
54
55
56
57
53
54

KEY

CUT

FILL

50

CUT AND FILL CIVIL ENGINEERING	DRAWN BY	FILE NO.	DRAWING 81

Plot the cut and fill areas for the level roadway at an elevation of 50 m, with a slope of 1:1 for the cut and 1.5:1 for the fill. Vertical Scale: 1/200. Crosshatch the cut and fill areas in opposing directions. Also crosshatch the rectangles under "KEY" below for identitification.

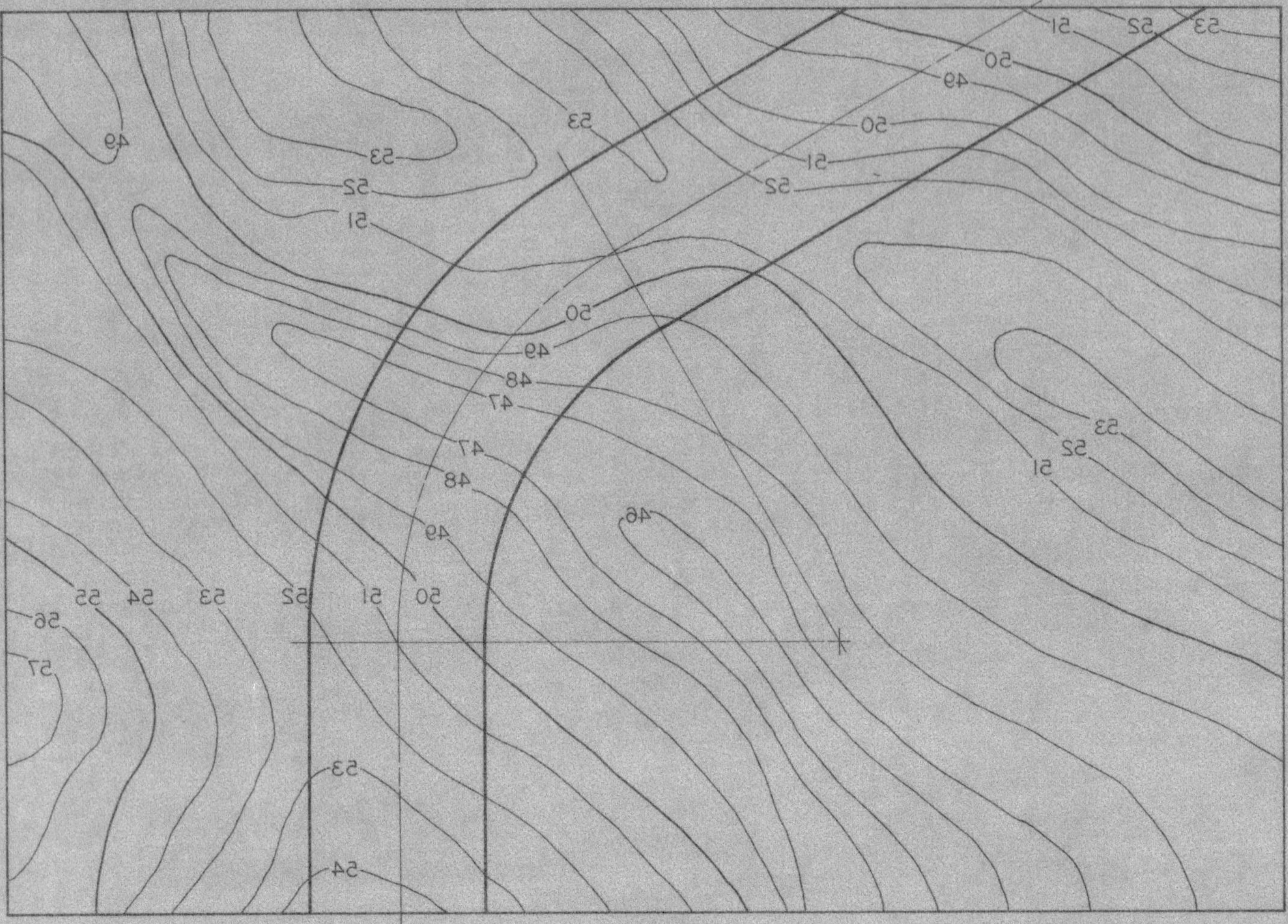

53
52
51
50
49
53
50
51
52
49
53
52
51
50
49
48
47
47
48
49
53
52
51
46
56
55
54
53
52
51
50
57
53
54

50

KEY

CUT

FILL

CIVIL ENGINEERING CUT AND FILL	DRAWN BY	FILE NO.	DRAWING 18

	DRAWN BY	FILE NO.	DRAWING

	DRAWN BY	FILE NO.	DRAWING

	DRAWN BY	FILE NO.	DRAWING